ANCIENT
ASTRONOMERS

S M I T H S O N I A N
EXPLORING THE ANCIENT WORLD
JEREMY A. SABLOFF, Editor

ANCIENT ASTRONOMERS

By ANTHONY F. AVENI

St. Remy Press • Montreal

Smithsonian Books • Washington, D.C.

EXPLORING THE ANCIENT WORLD
was produced by
ST. REMY PRESS

Publisher	Kenneth Winchester
President	Pierre Léveillé
Managing Editor	Carolyn Jackson
Managing Art Director	Diane Denoncourt
Production Manager	Michelle Turbide
Administrator	Natalie Watanabe

Staff for *ANCIENT ASTRONOMERS*

Editors	Michael Ballantyne
	Daniel McBain
Art Director	Philippe Arnoldi
Picture Editor	Christopher Jackson
Assistant Editor	Jennifer Meltzer
Researcher	Olga Dzatko
Photo Researcher	Anne duVivier
Photo Assistant	Geneviève Monette
Designer	Sara Grynspan
Illustrator	Robert Paquet
Systems Coordinator	Jean-Luc Roy
Administrative Assistant	Dominique Gagné
Indexer	Christine Jacobs
Proofreader	Judy Yelon

THE SMITHSONIAN INSTITUTION

Secretary	Robert McC. Adams
Assistant Secretary for External Affairs	Thomas E. Lovejoy
Director, Smithsonian Institution Press	Felix C. Lowe

SMITHSONIAN BOOKS

Editor-in-Chief	Patricia Gallagher
Senior Editor	Alexis Doster III
Editors	Amy Donovan
	Joe Goodwin
Assistant Editors	Brian D. Kennedy
	Sonia Reece
Senior Picture Editor	Frances C. Rowsell
Picture Editors	Carrie F. Bruns
	R. Jenny Takacs
Picture Research	V. Susan Guardado
Production Editor	Patricia Upchurch
Business Manager	Stephen J. Bergstrom

Library of Congress Cataloging-in-Publication Data
Aveni, Anthony F.
    Ancient Astronomers / by Anthony F. Aveni
        p. cm. — (Exploring the ancient world)
    Includes bibliographical references and index.
    ISBN 0-89599-037-7
    1. Astronomy, Prehistoric. 2. Astronomy, Ancient. 3. Indians—
        Astronomy I. Title.   II. Series.
    GN799.A8A84   1993
    520'.93—dc20   93-38463
                                                          CIP

Manufactured and printed in Canada.
First Edition

10 9 8 7 6 5 4 3 2 1

FRONT COVER PHOTO: *The rising sun strikes the spectacularly theatrical standing stones of Callanish on the Isle of Lewis, Outer Hebrides.*

BACK COVER PHOTO: *The tongues of panting dogs point to the ways of the stars on this 15th-century astrolabe.*

CONTENTS

EDITOR'S FOREWORD

People have charted the movements of the sun, moon, planets, and stars for millennia, and, as ongoing research enhances our understandings of ancient astronomical practices and expertise, our appreciation of the complexity and sophistication of the astronomical knowledge of ancient peoples has begun to grow as well. However, as Professor Anthony F. Aveni tellingly demonstrates in this volume, such astronomical know-how was linked so closely to theology that there is a temptation to deride it as astrology—rather than praise it as astronomy—and to underestimate its intricacy and the critical roles it played in past civilizations. As Professor Aveni strongly argues, though, such judgments can be misguided and too narrow.

Professor Anthony F. Aveni's strong scientific credentials and his judicious consideration of data add weight to his contention that ancient astronomy is worthy of both serious scholarly and public attention. Although some analyses of ancient astronomical phenomena appear more pseudo-scientific than scientific and seem to belong more appropriately on the pages of sensational tabloids than in books shelved next to weighty scholarly tomes in the "archaeology" section of libraries, Professor Aveni shows how investigations of prehistoric (as well as historic and modern) astronomical practices—particularly among non-Western peoples—can provide important new understandings of the nature of astronomy.

Professor Aveni is able to support his arguments through knowledge gained from long and detailed research all over the globe. He is internationally renowned for his studies of ancient astronomy. He has pioneered the combined use of archaeological and astronomical data, and is one of the founders of the new interdisciplinary area of specialization that has come to be called "archaeoastronomy." He holds a distinguished chair at Colgate University, where he is the Russell B. Colgate Professor of Astronomy and Anthropology. Professor Aveni's doctorate is from the University of Arizona. He has written voluminously on archaeoastronomy and is the author of *Maya City Planning and the Calendar* (with H. Hartung), *Empires of Time: Calendars, Clocks, and Culture*, and *The Sky in Mayan Literature*. He also is the editor of many books, including *Native American Astronomy*, *Skywatchers of Ancient Mexico*, *Archaeoastronomy in Precolumbian America*, and *New Directions in American Archaeoastronomy*.

This book fits admirably in the *Exploring the Ancient World* series because it provides an up-to-date assessment of the achievements of ancient astronomers by the leading scholar in the field who explicitly places these accomplishments in their cultural contexts. While the scope of the volume is impressively wide-ranging, Professor Aveni's carefully crafted prose and enlightening explanations, plus the well-selected illustrations, ensure that a potentially difficult and complicated subject is kept clear, lively, and understandable. Whether Professor Aveni is analyzing the Maya calendar or Polynesian navigation, he makes certain that readers are easily able to follow the discussion. Professor Aveni also takes care to dispel some of the common misconceptions that abound about ancient astronomical practices, as well as to guide the reader through the thicket of sensational claims about such phenomena as the Nazca lines in southern Peru or the great standing stones at Stonehenge in England and elsewhere. Readers of the chapters that follow will be both entertained and educated by Professor Aveni's insights into the astronomical practices and beliefs of the past.

Jeremy A. Sabloff
University of Pittsburgh

THE MOST ANCIENT PROFESSION

Twenty-five centuries ago, in an ancient Greek classroom, a teacher asks his students: "What shall we take up next? How about astronomy?" Whereupon a young scholar in the front row pipes up: "Good idea. It's practical knowledge. Good for learning about the seasons, how to navigate, keep time, and so on."

Socrates' bright student already knew that astronomy—the world's oldest science—was a practical business. From the islands of the Pacific to the jungles of Mesoamerica, people turned to low-tech, repeatable observations of the most reliable, precisely recurring events that can happen in nature: the daily and annual movement of the stars, the first gleam of sunrise, a sliver of a moon where none had been seen the night before, or the sudden appearance of the morning or evening star—all useful information for clocking daily affairs.

What Socrates' student did not express, however, and what I hope to demonstrate in this volume, is that behind the Maya written record of the movement of the planets, the Aztec alignment of their great sacrificial temple, the Babylonian cuneiform Venus Tablet of King Ammizaduga, and the Greek Athenian Tower of the Winds—behind each lies a heavily laden symbolic structure that we variously term cosmology, astral mythology, or (in our most suspicious posture toward things past) astrology. *Ancient Astronomers* aims to reconcile these two seemingly diverse approaches to understanding: the fantasy of myth on the one hand and, on the other, precise prediction in an astronomical timetable or a calendar. In the primal world of skywatching, myth and science meet at a crossroads.

The clear view of a dark, star-studded sky has become, for most of us, obscured by glaring city lights. Hopefully, the brief sky primers found throughout this book will bring the night sky back into focus. Only then will our celestial odyssey be able to map out the history of astronomy—from controversial Paleolithic scratchings on animal bone and Bronze-Age sky sightings, to the more easily decipherable astronomical documents of Indo-European cultures. We will pay special attention to the Babylonians and the Greeks, who gave the Western world the scientific foundation on which modern astronomy is built.

But there were other astronomical traditions as well, some of which tied in with the West, others which did not. We need to explore them, too, for together they embrace all of human intellectual endeavor and achievement.

Islam, for example, which spread over much of the Old World during the Dark Ages, coupled one of the highest forms of mathematical reasoning with superior instrumentation to surpass the astronomy of its Aegean predecessors. And in Asia, a bureaucratic-style tradition of recording and predicting sky events even farther back into the past failed to fuse completely with that of the West, though astronomical knowledge was traded. Across the Atlantic, New World civilizations created entirely distinct celestial cycles and invented their own ingenious astral metaphors. As early as the sixth century, the Maya developed the art of mathematical celestial prediction on a par with that of the Old World. And by the 15th century, the Incas of Peru had created a sophisticated cosmology that integrated ideas of kinship and astronomy. In the middle of what is now the continental United States, indigenous people—once erroneously thought to be the Lost Tribes of Israel—dedicated some of the largest monuments in the world to astronomical reckoning and sky worship. The ancient record from Africa and Oceania is a good deal more sparse, yet we can piece together suggestions that there, too, people had invented calendars and encoded their myths in the constellations they envisioned. They had built their households round the stars and they had sailed by them.

While there are common denominators, there are also significant differences between all these human encounters with the sky. By sampling the rich record of our ancestors from all over the world, we will discover that the practice of ancient astronomy was just as diverse as the religions, economies, agricultures, and social customs of the peoples who created it. We will see how each culture imposed its own interpretation and structure upon the world of nature—how each society organized astronomical knowledge in ways that suited its needs, integrating it with other categories of knowledge, such as religion or astrology, which do not necessarily enter into scientific astronomy today. To understand why astronomy was so important to these ancient peoples, we must penetrate the cultural contexts in which it was embedded.

Since the beginning of time we have looked skyward in our attempt to understand the nature of things and the relationship of celestial produced its own astronomers whose ingenuity and imagination still impress us today.

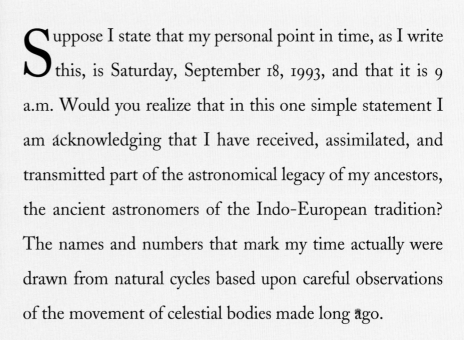

1

A SKY
FOR EVERYONE

Suppose I state that my personal point in time, as I write this, is Saturday, September 18, 1993, and that it is 9 a.m. Would you realize that in this one simple statement I am acknowledging that I have received, assimilated, and transmitted part of the astronomical legacy of my ancestors, the ancient astronomers of the Indo-European tradition? The names and numbers that mark my time actually were drawn from natural cycles based upon careful observations of the movement of celestial bodies made long ago.

happenings to human endeavor. Every culture

To prove my case, let me trace out each wheel of time that grounds me in the present. Nineteen ninety-three represents the largest cycle. It tells me that the sun has migrated all the way around the constellations of the zodiac one thousand, nine hundred, and ninety-three times since the putative birth of the Christian Savior. (Debate continues about precisely when Christ was born; and although astronomers have correctly tallied the number of days in each yearly cycle at 365, historians in the Middle Ages still managed to set the starting point of the count incorrectly.)

The name "September" refers to one of 12 months—a contraction of "moonths" (which hints at the nature of their origin). And so the September page of my wall calendar will remain unturned for approximately as long as it takes the moon in the sky to pass through its full cycle of phases, from crescent to quarter to gibbous to full, then in reverse until it disappears in the morning sky, only to reappear in the western twilight as first crescent. The number 18 after the word September once told my ancestors, who had no such calendar to consult, that two less than the number of days that they could count on all their fingers and toes had passed since they had first glimpsed the thin crescent moon that had most recently appeared in the west at dusk. They also noticed that this time period (called a lunation) measured the interval between successive menstruations in the human female.

During the first century B.C., Roman politics, rather than astronomy, played a major role in distorting the measure of solar and lunar time. Because a whole number of months (strictly measuring the lunar cycle) did not fit exactly into the year of the seasons, Julius Caesar artificially lengthened the number of days making up each calendar page, thus stretching the month beyond the reality of nature. He was careful to include his own month (July) among the lengthier ones! A generation later, Emperor Augustus made certain that no month, not even that of the more famous Caesar who preceded him, would have more days than the time unit he named after himself. But why, as the Latin etymology implies, is September not the *seventh* month of the cycle? The answer is that we inherited our year from the Roman Empire, where they began the count with March, the month containing the spring equinox. This is the day when the sun rises precisely in the east and sets exactly in the west—when day and night are equal in length. Another Roman innovation was more practical in nature. The cycle of the seasons, approximately 365¼ days, was a bit longer than the yearly 365-day count of the days, so the Romans *intercalated*; that is, they added extra days to the calendar— one every four years—so that time's canon would not get out of joint with the perceptible rhythm of the skies.

If month stands for the moon and year for the sun, then what is the significance of the division of weeks and days? "Saturday," or "Saturn's Day," may give us a clue, but even on a time-related question as basic as this one we run into controversy. Modern chronologists say that there is a natural biological cycle in all humans that oscillates over a seven-day period. But in many cultures the week is also a convenient market cycle. It is roughly the length of time it would take a family to pick and sort a batch of home-grown produce, take it to a market place, sell or trade, perhaps worship, celebrate, and feast with a wider communal circle in a place of assembly, and finally, to return to the hearth to undertake other chores while another load, about all one could carry, would ripen and be ready for another distribution cycle.

Our ancestors—who recognized the lunar rhythm in the bodily cycle of women—may have witnessed a parallel between the market cycle and the movement of celestial bodies. The seven-day week is approximately equivalent to a quarter of the lunar phase cycle; it is, for example, about the length of time it takes the moon to change from first quarter phase to full moon, or from full moon to last quarter. Furthermore, to bring religion into the picture, seven is also the number of visible planetary deities in the Babylonian celestial Pantheon: Saturn, Jupiter, Mars, Sun, Venus, Mercury, Moon, here arranged in order according to the diminishing size of their presumed orbits about the earth, long presumed the fixed center of the universe. (I have disguised the Babylonian deities in the more familiar nomenclature of their Roman equivalents.) We will learn in later chapters that the ancients connected biological, astronomical, and economic cycles together—along with religious cosmology—in just this way.

The hour of 9 a.m. turns out to be one of 24 major segments of another cycle—the rather short solar cycle that we call the day. The hour of the day was once indicated by human gesture—a simple raising of the hand toward

The Babylonians believed the earth was the fixed center of a water-bound universe and that the seven visible planetary deities were arranged around it according to the diminishing size of their presumed orbits. They are shown here using their more familiar Roman names.

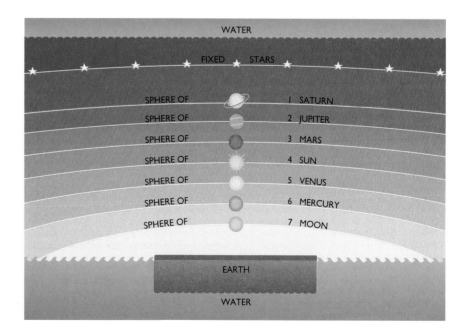

The names of the days in English have mixed origins. They were inherited from Teutonic tribes, but the Teutons had borrowed some names from the Romans. Thus, Sunday and Monday, honoring the sun and moon are Roman and so is Saturday—named for the Roman god Saturn. The remaining days are more familiar to English-speakers in the guise of the Teutonic gods (or equivalents), hence Tiw (the god of law) for Tuesday, Woden (principal Teutonic god) for Wednesday, Thor (after the god of war) for Thursday, and Fria (goddess of love) for Friday.

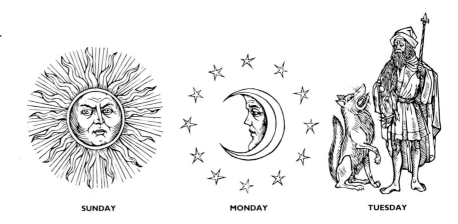

SUNDAY　　　　　MONDAY　　　　　TUESDAY

the sun to signal the hour for prayer, for sustenance, or for bringing in the cattle from the pasture. Today, the little hand on the face of the clock has replaced the human timekeeper's extended arms, but it still does its best to point to the sun in the sky. We designate three hours past the meridian—the hour of noon being when both hands point straight upward—as 3 p.m. (post meridiem, actually), and three hours before (ante), as 9 a.m. Though all hints of archaic human gesture vanish on a digital watch, its hours, minutes, and seconds continue to betray the ancient astronomers' attempts to link the large solar cycle with the small. Some clever time-manager in ancient Babylon partitioned the diurnal (daily) movement of the sun into units that would be divisible into "the sun of the year." Thus we have, in sexagesimal notation (based on the number 60), 60 minutes in an hour, 24 hours in a day, and approximately 360 days in an annual cycle of time. It is no coincidence that 360 is also the number of divisions in a full circle of space, what we have come to call degrees; we can think of each degree as a measure of advancement that marks the journey of the sun across the sky.

Priests, undoubtedly in close alliance with their astronomers, assigned dominion of each of the day's 24 hours to one of the planets, running through the list of seven luminaries from the top of the hierarchy down, and naming each day after the god who ruled the first hour. Thus we once called the first day of the week Saturday—or Saturn's day—and the second one Sunday. If you count off by threes (the remainder obtained after dividing seven into 24) in the descending order of planetary orbits shown on page 13, you will discover the hidden logic behind the order of our modern day names—though you will need a little help from other invasive mythologies to arrive at the names as they are written in English on today's calendar. For example, Fria is the Nordic equivalent of the Roman love goddess Venus, the regent of the last day of the week.

WEDNESDAY

THURSDAY

FRIDAY

SATURDAY

My current point in time, then, masks a hidden bequest from my ancestors. Our clocks and calendars harbor a wealth of embedded celestial knowledge that we tend to employ rather indifferently, and which we apply to quite different ends from those of the ancient timekeepers who created it. We reorganize archaic knowledge into patterns that suit our own common sense. We secure it not by skywatching, but by building mechanical models consisting of moving parts that oscillate from one extreme to another, then back again to the original state—the pendulum, the balance wheel, the microscopic crystal of a timepiece—these devices we use to "watch" nature. We take our cues from nature and we reprocess the measure of astronomical time, padding or shaving the hours of the day and the days of the year—even lengthening an occasional weekend to suit our modern needs or setting the clock back an hour in the fall or forward an hour in the spring to "save" daylight. Recently we added an extra second to the clock (June 30, 1993) to keep the yearly calendar on schedule).

By studying each segment of a given moment such as Saturday, September 18, 1993, we can illustrate that what we see in the sky was once perceived as knowledge of a different kind. We think of astronomy today as the study of the universe of matter and energy that lies above the earth's atmosphere. Such a definition can separate the sky from you and me, from the human spirit, from the concerns of daily life. However, politics and biology, history and economics, religion and mythology have conspired together in the process that leads to that tiny block almost two-thirds of the way down and to the right on page 9 of my 1993 wall calendar.

You will also appreciate that the cycles of sun and moon, of planets and stars that come down to us in that date are a kaleidoscopic jumble of inherited rhythms and periods passed on from one generation—as well as from one culture—to another. Some temporal facets have fallen by the wayside, others have been chipped and remolded—both with and without reason—to adapt to an ever-changing common sense.

What, then, is worth knowing about the world's first astronomers? Certainly it is worthwhile to examine the nature of astronomy's roots—the fitting together of each fragment that makes up the pattern we see today through nature's kaleidoscope. But we should also pay attention to what that pattern looked like at the last quarter or half turn of the kaleidoscope of human history. What made up a culture's body of sky knowledge? Knowledge is both information and process, however. So we must also ask questions about the hand that turned the kaleidoscopic tube of time. What motivated human nature to turn it this way, or that, slowly or quickly? How was astronomical knowledge acquired and what purposes did it serve? What were the driving forces that led ancient civilizations all over the world to look at the sky to seek order and structure in their lives? How was astronomical knowledge expressed in architecture, oral poetry, conventional writing? What part did astronomy play in the development of a culture's world view, its ideology, its way of relating individual and society to nature and the transcendent? We must raise these questions not simply in the context of our particular scientific tradition, but also in light of all the conceivable ways through which our ancestors have tried to understand the cosmos.

Stripped to its essentials, then, the subject of ancient astronomy is really about human inquiry into the nature of things—in particular, things celestial. Given the fact that, basically, the same celestial images glided over the heads of Polynesian and Babylonian, Mayan and Egyptian, how did these various organized societies, living under a canopy of flashing and moving lights, interact with what they saw? The answers will be both surprising and simple. They did it in wonderfully diverse and imaginative ways. So diverse that we will come away with a fuller appreciation of the ingenuity of the ancient astronomers. And so imaginative that some of us will conclude our voyage through their lost astronomies with a feeling of envy rather than the superiority we too often adopt toward the past. While most of us are cozy under the umbrella of beliefs and conveniences that characterize the progressive present, we may nevertheless come to regret having lost our ancestors' ability to carry on a dialogue with nature's forces.

Before looking more closely at how people all over the world saw the sky, a little philosophical perspective is in order. Modern astronomy organizes celestial knowledge within a gargantuan framework of imperceptibly long time intervals and inexpressibly wide spatial extensions. Like all branches of modern science, it bases its faith in the belief that a unity exists among nature's forces; for example, we believe that gravity behaves exactly the same way whether it acts upon an apple falling to the ground or a star orbiting the core of a galaxy millions of light years beyond the range of our most powerful telescopes. If our first article of faith is unity, our second is progress. We believe that the totality of knowledge we can perceive about the universe will lead us toward the truth about what is really going on. But truth for the modern scientist is never ultimately and unquestionably attainable; it is only reachable approximately and by degrees (asymptotically, as the scientist would say). Modern astronomical science is destined to create not only the theory or model that best yields predictions that can be tested empirically, but also—and herein lies the rub—theories or models capable of being *disproved*. This is our credo, our faith.

This built-in stress—the risk of mistake, incorrectness, and uncertainty—hangs over the head of the modern scientist in much the same way that the vagaries of a Maya Itzam-Na or a Greek Zeus (to mention just two of many sky gods) loomed over those of our mortal counterparts in the New and Old Worlds. Our uncertainty about the way the universe functions is at least as great (if only subconsciously) as that of our ancestors.

And now we will attempt to watch the night sky—not as we see it in the overlighted urban centers that about half of us inhabit, but out in an open field in the country, far away from all artificial luminous distractions. This is the setting in which our predecessors began the watch on nature that culminated in a multitude of astral expressions that survive in the archaeological and epigraphic record—including our own calendar. By attempting to see the sky through their eyes we will be better equipped to appreciate the variety of symbolic forms our forebears assigned to it.

We live in an ordered world of observable natural phenomena whose constancy and consistency are typified by events in the sky, whether the sky event be a sunset, a close encounter between two planetary wanderers, an eclipse, or a shower of shooting stars. Nowhere else in nature—not in the comings and goings of the birds, the blossoming of trees, nor the arrival of the rains—do we find a more reliable environmental reality in which to frame the drama of life than the celestial backdrop.

How do the stars move from hour to hour and from night to night? Which way do the sunsets shift on the horizon? What changes does the moon undergo? Where can we expect to see it appear and disappear each month? What is the zodiac and why was it so important to the ancients? How do the sun, moon, and planets advance across this great celestial roadway? When will they pass one another again or slip through that curious periodic backward motion the Greeks called retrograde? How often do eclipses occur and comets appear, and why were they so universally feared and dreaded? Because all of these phenomena were the grist for the many mills of astronomy in antiquity, we must become familiar with each of them as we encounter the desires and needs of the ancient skywatchers who we believe actually predicted their recurrence. We begin with the most ancient astronomers of all—those who made sky forecasts without the aid of any written record.

The moon shines dramatically at the heart of the sacred stone circle at Callanish on the Isle of Lewis in the Scottish Hebrides. People assembled captured in this position, an occurrence that takes place every 18.6 years when the moon follows its extreme southerly path.

2

THE UNWRITTEN RECORD

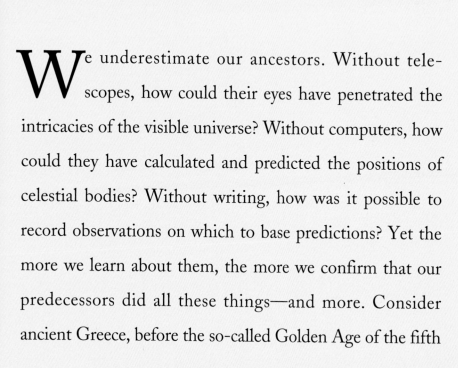

here more than 4000 years ago to spot the moon

We underestimate our ancestors. Without telescopes, how could their eyes have penetrated the intricacies of the visible universe? Without computers, how could they have calculated and predicted the positions of celestial bodies? Without writing, how was it possible to record observations on which to base predictions? Yet the more we learn about them, the more we confirm that our predecessors did all these things—and more. Consider ancient Greece, before the so-called Golden Age of the fifth

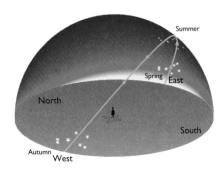

This illustration charts the progress of the Pleiades from their first appearance in the eastern sky in spring, when Hesiod in his poem recommends the start of spring planting, through summer, when they are at their highest point in the sky, and finally to autumn, when they set on the western horizon. In each instance the pre-dawn glow of sunrise is depicted in the east.

century B.C. The Parthenon had not yet been built, nor had the Persian Wars been fought. We call it the prehistoric period because there is no recorded documentation of what really happened. Yet astronomy was alive and well even then. In 800 B.C., the northern Peloponnesus was farmland unfit for the faint-hearted. There, a hard-bitten farmer named Hesiod once composed a poem. Actually he recited it, though a later written form comes down to us. A good deal shorter than *The Iliad* or *The Odyssey*, yet composed in the same enquiring spirit, its 828-line description of farming techniques constitutes a moral lesson about how to outwit fickle nature. If we read Hesiod's *Works and Days* carefully and penetratingly, we find answers to many of the astronomical questions we asked above. Hesiod's eyes perceived as well as our own how the stars move across the sky. In verse he calculates and predicts the course of nature by carefully following the stars. And he keeps track of everything in his own mind—that wonderful device whose capacity for remembering has all but atrophied in the age of artificial memory devices such as the computer.

Hesiod tells us when to plant and harvest, not by the printed calendar, but instead according to a schedule signaled, for example, by the arrival and departure of a star group like the Pleiades.

> When you notice the daughters of Atlas, the Pleiades, rising, start on
> your reaping, and on your sowing when they are setting. They are hidden
> from your view for a period of forty full days, both night and day, but
> then once again, as the year moves round, they reappear at the time for
> you to be sharpening your sickle.

Even today, the notion that the Pleiades are a harbinger of rain is not lost. The modern English poet A. E. Housman wrote in his 1936 *More Poems*:

> The weeping Pleiads wester,
> And the moon is under seas;
> From bourn to bourn of midnight
> Far sighs the rainy breeze:
> It sighs from a lost country
> To a land I have not known
> The weeping Pleiads wester,
> And I lie down alone.

The illustration above demonstrates what Greek skies would have looked like when the Pleiades made their first annual appearance in the pre-dawn eastern sky after having been hidden by the sun for several weeks (what astronomers call the heliacal rising). Now is the time to put your seed in the ground, says the farmer-poet. The actual date extrapolated to the latitude and time period works out to be late May. From that point forward in the seasonal year, the Pleiades appear higher and higher in the pre-dawn sky. By the middle of the summer they traverse the sky like a clock's hour hand for most of the

20

The Peleponnesian peninsula was home to Hesiod, the Greek farmer-poet, who in the ninth century wrote about when to plant the seed, when to harvest, and when to make the wine based on his observation of the stars.

night. Finally, the conspicuous little cluster arrives at the opposite horizon and faces the rising sun. When this happens, says Hesiod, harvest your crop.

Hesiod even has a celestial recipe for wine-making. It begins with a tricky mnemonic that commences with the occurrence of a double celestial event:

> But when the stars of Orion and Sirios have climbed up into midheaven and rosy-fingered Dawn is facing Arkturos, then, Perses, pluck and bring home all your clusters of grapes. Set them to dry in the heat of the sun for ten days and nights, and in the shade for five days, and then on the sixth day draw off the blessings of glad Dionysus into your jars.

The lines are poetic in the original Greek—but they go beyond mere lyricism. In fact, they are laden with primal science. For example, when do the red streaks of cirrus cloud arch out to touch the orange star Arcturus as it first stands above the pre-dawn horizon? Or, in astronomers' parlance, when does Arcturus rise heliacally? It is at the same time that, turning southward, you will see Sirius chasing Orion across the meridian. Pick the grapes then, says Hesiod. Then count forward all the fingers on both hands; put the grapes in the shade; count another hand's fingers, plus one—then press the grapes.

Hesiod's oral farmer's calendar is jam-packed with naked-eye astronomy. He used the sky as an index of the appropriate time to plant herbs, to winnow and store grain, as well as to anticipate the advent of garden pests and the arrivals and departures of birds such as the crane, the swallow, and the cuckoo. He even offers advice on the best time to go sailing. The information in the *Works'* calendar is not a storehouse of astronomical knowledge compiled for its own sake. Instead, it is concerned directly with the everyday struggle for human existence.

Hesiod basically is concerned with saving his farm. In his harsh environment, rural life depended on watching the stars, and so he sings his song to educate his wayward brother Perses, who—unless carefully instructed by his more responsible (not to say nit-picking) brother—is liable to mismanage the family legacy. The poem not only tells how to pick up environmental cues, it also reveals the perceived interconnectedness of nature's parts. Hesiod ties human activity—pruning and digging, hauling and pressing—to environmental changes, such as weather, the seasons, and animal and plant behavior. He frames all of the events he describes in the context of nature's most dependable timepiece—the sky.

Is it any wonder, given this practical sort of calendar-keeping, that celestial bodies came to be thought to directly influence the behavior of things here below? Sky objects became deified. They were worshipped, revered for the good that they brought; they were offered sacrifice as a means of averting misfortune. While we may not think of Hesiod's poetry as real science, we shall see in the next chapter that his seminal efforts mark a first step, in the Middle East and the Aegean, toward the implantation of the taproots of modern astronomy.

How far back can we trace this unwritten form of record-keeping? Before Hesiod, we have no knowledge of hand-me-down fireside tales about the sky. But we do have some tantalizing imagery engraved in bones and stones—markings and orientations that, if some interpreters are correct, could catapult the time band of astronomical record-keeping backward several millennia before Hesiod's era.

A carved piece of bone—part of an eagle's wing—recovered in a cave in central France early in the 20th century has been dated by archaeologists to 30,000 to 32,000 B.C. One of many such fragments with stroke and point marks, it rested in the dusty shelves of France's Musée des Antiquités Nationales in St. Germain-en-Laye until Peabody Museum researcher and writer Alexander Marshack carefully examined it under a microscope in 1965.

The user seems to have held the bone in the palm of one hand and used a twisting motion to gouge out each mark with a stylus or other pointed tool. Upon reaching one end of the bone, he or she turned it 180 degrees in the hand and created a linear pattern that runs in the opposite direction. The twist of the comma-shaped marks visible under the eye of the microscope proves that they are ordered sequentially. But for what purpose? What had previously been thought the marks of a tool sharpener, Marshack interpreted as a time-factored notational record of the lunar-phase cycle, each point standing for one day. He also noticed a single continuous, sinuous pattern that linked all the dots. It was as if all the gouged-out points could be connected by one sweeping curve. Counting and grouping the dots, Marshack detected clumps of 15s and 30s that corresponded to each of the changes of direction of notation—intervals that fit major changes in the lunar phase cycle. To date, the bone carver is the first person thought to have kept a record of what transpired in the heavens.

The obvious question arises: Why would a hypothetical Paleolithic calendar-keeper have created such a lunar record? What was he or she recording, and for what purpose? To answer that question, we need to know how the moon moves. The moon, like the sun and the planets, treks along from west to east against the stars, moving fastest of all about the zodiac. It takes only a month. But the moon also has one property exhibited by no other of its sky partners: It changes its face. First visible as a thin crescent, its horns point away from the setting sun over which it first can be briefly sighted. Babylonian astronomers once perched atop the roofs of their temples and stood watch for it to signal the start of a new month. Next night it would appear at dusk slightly higher in the sky; it would be higher still on the third night as it waxed to quarter; then on to gibbous; and finally full phase. By that time, 14 or 15 days after it has first been sighted, the full moon rises in the east, directly opposite the setting sun, with the observer standing on a direct line between the two. Did Paleolithic cave dwellers know what later Babylonians and Mayans would determine—that every fifth or sixth of these full moons was liable to have its face turn a ruddy color in total eclipse?

A carved piece of bone from an eagle's wing, found in a cave in France's Dordogne Valley, has been dated by archaeologists to 30,000 to 32,000 B.C. The markings, originally thought to have been made by a tool sharpener, were interpreted in the 1970s to be the notational record of the lunar phase cycle— the first evidence of a human record of celestial events. Peabody Museum researcher Alexander Marshack linked all the dots on the bone, and theorized that the turning points might coincide with new and full phases of the moon.

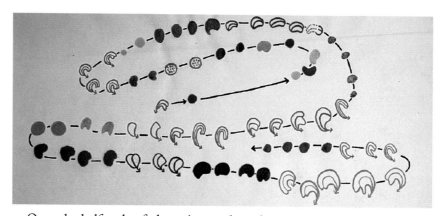

Once the half cycle of phases is complete, the moon proceeds to wane from gibbous back through quarter, now more accessible to the daytime observer. Finally, only a thin crescent remains, positioned just above the rising sun. The lunar life cycle parallels the life story of the man depicted in profile on its face. Like a young king who accedes to the throne, the man in the moon's career waxes to brilliance and success; his rule is at the peak of effulgence when his face is full. Then the evil dragon of darkness nibbles away at his countenance, seeking to conquer and destroy him as he had so dethroned his father, the old king. But the next of the moon god's generation reappears from his father's ashes to challenge and conquer the forces of evil all over again.

We can imagine that sitting by the fire at the mouth of the cave at twilight, our Paleolithic recorder of moontime might have noticed the first thin crescent moon low in the west—his first mark. The next evening the observer saw a thicker crescent, slightly higher in the sky—mark two. When the seventh mark was made, a quarter moon would have stood high on the meridian at sunset. By now the moon began to provide faint light for a few hours after sunset. The gibbous phase brought a still longer duration of yet brighter moonlight. By the 14th or 15th day of the cycle, the full moon—appearing opposite the setting sun—remained in the sky all night long, allowing some limited nighttime activity, like hunting nocturnal animals or fishing. By 21 days, the moon

23

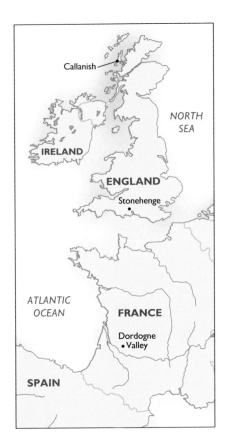

Two major megaliths are located in Britain—Callanish, in the Hebrides, and Stonehenge, on Salisbury Plain. The Dordogne valley in France is where the first known lunar recording was found, on a piece of eagle bone.

already would have begun to regress in its phases. When it reached last quarter, it rose at about midnight and could still be seen in the daytime sky after sunrise. By the 28th or 29th day, the intensity of lunar light would diminish to a waning crescent, last spotted heralding dawn's appearance. Its horns would point skyward as it lingered over the sun's pre-dawn glow; and then it would disappear for one or two days (depending on the season of the year). Once again the night sky returned to total, impenetrable darkness.

Perhaps our cave dweller even noted the moon's place of disappearance or reappearance on the horizon. We have no evidence of this, but we do know that the Ngas of contemporary Nigeria are concerned with such sightings. Perhaps, as with the Ngas, the left- or right-hand tilt of the crescent was a sign of rain.

As in Hesiod's much later time, there is no question that Paleolithic and Neolithic cave dwellers, whether sedentary, semi-sedentary, or even nomadic, would have had good reason to engage in moon-watching. It is likely that the women of the Paleolithic cave culture in southwestern France followed the moon as a periodic light source, and also that the women of that culture knew that the menstrual cycle was synchronized with the lunar phases and that they could even count the time to child-bearing by tallying a particular number of visible lunar disks. Lunar changes were obvious harbingers of things to come; they gave society a lease on the future. Lunar scheduling offered an adaptive advantage, a measure of order amidst what otherwise might seem random or chaotic.

How long can we remain on this side of the river before the flood season will hinder our crossing back? When can we anticipate that the fruit on the vine will ripen? Will there be enough time to get our supplies into winter storage and move ahead to the hunting ground? An intimate knowledge of nature's timing devices demands a good eye, as well as a means of remembering and passing on collected information. Those old bones in the museum case may constitute all that remains of this sensitive archaic vision and insight into the first schedule of life's activity sheet. But there were others:

> This island...is situated in the north and is inhabited by the Hyperboreans...And there is also on the island both a magnificent sacred precinct of Apollo and a notable temple which is adorned with many votive offerings and is spherical in shape...They say also that the moon, as viewed from this island, appears to be but a little distance from the earth and to have upon it prominences, like those of the earth, which are visible to the eye. The account is also given that the god visits the island every nineteen years, the period in which the return of the stars to the same place in the heavens is accomplished;...(Diodorus of Sicily, Book II 47:1-5, 5-48)

In midwinter on Salisbury Plain, in southern England, it gets dark rather early. Cold and windswept, this is an inhospitable place. Nonetheless, we can be sure that our ancestors once assembled there on Midwinter's Night (December 21) 5000 years ago at the megalithic monument known as Stonehenge (it means hanging stones). On that occasion several allied tribes

Stonehenge, on Salisbury Plain in southern England, was the gathering place as long as 5000 years ago for people who came to worship their gods of nature, the sun and moon. The huge stones are positioned so that the observer, standing at the circle's center, can capture the sun's disk in the gateway on the first day of summer. Other alignments to the sun and moon were witnessed through the great trilithon (three-stone) archways, one of which is seen still standing to the left of center.

gathered within the 427-foot- (130-meter-) diameter circular ditch-and-bank enclosure, the so-called "spherical temple" that the Greek historian Diodorus wrote about in the first century B.C. They stood among erect, people-like stones—special worked columns that had been quarried, shaped, and carted by sledge all the way from the Prescelly Mountains in Wales, a distance of 135 miles (217 km) as the crow flies. They came to this place to worship their gods of nature, the sun (Diodorus calls him Apollo), and the moon. Perhaps they also traded goods in a market and celebrated a communal festival. On that night they also watched the full moon rise on, or close to, Stonehenge's main accessway, and they likely realized that when it rose precisely in the 16-foot- (5-meter-) high stone gateway 87 yards (80 meters) northeast of the center of the circle, the next full moon would be eclipsed. But even if no eclipse took place, that special midwinter full moon rising opposite the setting sun would provide ample light for a night-long ceremony to honor the attending gods.

Today, only the right-hand standing stone of the 5000-year-old gate remains—it is called the Heel Stone—but the rising midwinter full moon still keeps its ancient appointment. It completes a cycle of arrivals and departures along the east horizon every 19 years, just as the Greek historian tells us.

In the opposite season of the year, on Midsummer's Day, or June solstice, the sun itself rises over the Heel Stone; thus it marks out one of its extremes or standstill points on the horizon. Like a pendulum, the sun would roam over the full extent of its horizon track, returning to its starting point in 365 days. Over the course of the seasons the sun shifts its rising and setting position along the horizon. It appears farther to the north in June and farther to the south in December (the solstices). In mid-March and mid-September (the equinoxes) it ascends and descends over the horizon precisely at right angles to the north-south direction. Following the sun at horizon was one way native people of the western United States as well as the high Andes charted out their agricultural cycles. But in ancient Egypt as well as ancient Mexico, the first astronomers marked out the year in still another way. They tracked the annual path of the sun through the zodiac. This is a band of 12 (in some cases 13) constellations that extends all the way around the sky. Many people recognized that with each full cycle of lunar phases one constellation replaced another over the place where the sun had set in the west. The constellation of Sagittarius replaces Scorpio, which replaces Libra during the rainy season in the ancient Fertile Crescent of the Middle East. Little wonder that the next three constellations have aquatic-sounding names—Capricornus (the Sea Goat), Aquarius (the Water Bearer), and Pisces (the Fish).

Though modern astronomers look up, measuring celestial coordinates in a system that turns with the sky and charting planets by giving their positions along the zodiac, ancient cultures kept track of time predominantly by horizon astronomy. One advantage to astronomical horizon watching is clear. All it takes is a marker—a natural peak or valley or a marker-stone deliberately placed in a strategic location—and then an observation of sun or moon on the marker constitutes a clock. No numbers, no notation are necessary.

Just as we have unearthed other prehistoric bones with notations inscribed upon them, so too have we stumbled over other standing stones—megalithic observatories occur in England, Scotland, and Ireland, Northern France, perhaps even in Malta, Morocco, and Romania—not all built by the same culture to be sure, but nevertheless preserving evidence that they once functioned in part to register sky events.

For us, architecture has a singular function: Joe's Garage is only a garage. We do not think of it as a place where men gather on Monday morning to rehash the weekend football games—even though that is what happens. And the local church is considered only a place of worship, even if Bingo games are played in its basement on Friday nights. To understand the ancient perspective of the sky we need to get away from the notion that their observatories must be like ours. We shouldn't fall into the trap of garbing our forebears in our rather restrictive scientific attire, of endowing them with the same rationale to philosophize, to hypothesize, to practice experimental science for the same set of reasons as we. Why deceive ourselves when we have before us evidence of a diverse and wonderful world of ancient human imagination to explore? Stonehenge and its sister megalithic complexes were erected not just to gain access to the sky and to set up an orientation calendar, but also to bring seminomadic people together, to conduct rites, to see the sky gods in all the right places. They were spiritual, political, nurturing—all categories of human behavior that included an understanding of astronomy. So we can think of

Our 12 zodiac signs were devised by the ancient Greeks and named after the band of constellations through which the sun, moon, and planets pass. During a year the path of the sun across the heavens moves through one constellation sign per month. In this medieval French illuminated manuscript, the sun is shown in the constellations of Aquarius and Pisces over the depiction of winter, and in Gemini and Cancer over that of summer.

Stonehenge as a place of social gathering, of religious assembly, as a cultic center, an economic center, as a place of fortified habitations, a celestial temple, an observatory. All of these definitions crosscut one another, some perhaps being stressed more at one time than another. The great achievement of Stonehenge is that the genius of its inventors encapsulated all of these functions in a single monument.

But Stonehenge was not built in a day; in fact, it was constructed, deconstructed, and reconstructed over a period of some 2000 years. First came the ditch and bank structure containing the 56 Aubrey Holes (the evenly spaced, chalk-filled holes arranged in a circle about the periphery) and the solstice gateway—what archaeologists call Stonehenge 1 (about 3000 B.C.). Then it

27

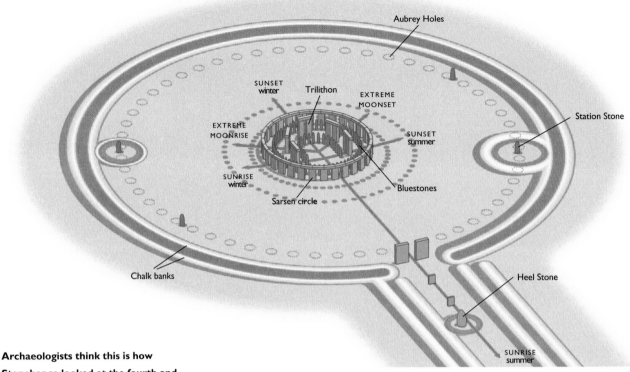

Aubrey Holes

SUNSET
winter

Trilithon

EXTREME
MOONSET

EXTREME
MOONRISE

Station Stone

SUNSET
summer

SUNRISE
winter

Bluestones

Sarsen circle

Chalk banks

Heel Stone

SUNRISE
summer

Archaeologists think this is how Stonehenge looked at the fourth and final stage of its construction—some 1200 years after it was started in about 2750 B.C.

may have been a simple place of assembly with a built-in solar timer to tell people the most opportune time to gather. Imagine how impressive the rites to the sun god would have appeared with the morning sun shining down the accessway. Seen by whom? The archaeological record suggests that Stonehenge was erected by several extended families, each consisting of approximately 50-100 people. They were colonists who came in search of a good place to cultivate grain and later to raise cattle. There is evidence that their descendants and other newcomers may have mismanaged the environment to the point of crisis, denuding it of trees and offsetting the chemical balance in the soil.

By 2500 B.C., the population had increased, tending more toward pastoralism and intensive farming. They built and lived in huge roundhouses, erected chambered tombs and mausolea. Some members of this society hunted on the high moors; others mined. Scholars agree that the culture that erected the huge trilithons, which now added lunar to solar extremes as timing devices at the horizon, must have been highly ranked, with specialized groups assigned individual roles in the project. Watching and marking the moving sun and moon with the degree of detail turned out in the alignments would have constituted a full-time job. You cannot lay out all of those orientations in a frequently cloud-bound environment without a lifetime or more of careful observation and impermanent staking. We can think of these early astronomers as engineers of a sort, who spent a good deal of their time working at the task of laying out the alignments, correcting and recorrecting, updating and improving their precision.

Why is Stonehenge a circular monument? It is not difficult to imagine that the idea for the plan of Stonehenge, or neighboring Avebury, derived from the conversion of a communal wooden roundhouse into a more permanent ceremonial center. The process may have resembled the enshrinement of an ancestor after death, except that in this case it was the symbol of the family, the domestic dwelling place that was given permanence in the ring of very large, round, standing boulders called sarsens. The tradition of the woodworking capacity of these ancient people may have been carried on in the linteled structure of the sarsen ring and trilithons, and in the vertical arrangement of standing stones that resemble the timber-settings in the domestic roundhouse, according to archaeologist Aubrey Burl. We have no record of who conceived this way of monumentalizing certain aspects of their society within the single-access, astronomically oriented ring structure that is Stonehenge, but it is clear that whoever they were, they had built upon the ideas of their predecessors who had constructed the ditch-and-bank structure more than 1000 years earlier. We might compare them to the architects of the cathedrals of medieval Europe, men whose awe-inspiring structures were monuments to the glory of God, like those still standing at Chartres or Cologne, Amiens, or Rouen.

The Gothic cathedral of Chartres in northern France was erected to the glory of God in the 13th century in much the same way as earlier structures had been built to honor various sky deities in prehistoric times.

An astronomical tradition was carried on at Stonehenge as well. There can be no doubt that it was a sun temple from its very inception. Without question it was deliberately positioned so that the viewer, standing at the center, could capture the sun's disk on the first day of summer in the Heel Stone gateway when the sun reached its northern standstill. Stake holes adjacent to the Heel Stone once held wooden posts that marked the moon at its major 19-year standstill. Within the ditch and bank, even in its earliest phases, lay the four station stones. Alignments among these stones marked out the maximal stretch along the horizon between the sun at one of its standstills and the moon at its opposite. A 90-degree difference between them occurs only in the latitude of Stonehenge. Whether the 56 Aubrey Holes within the ditch and bank may have been used as an archaic computer to tally intervals between the dynamic encounter of these two celestial luminaries in the form of eclipses, as astronomer Gerald Hawkins has suggested, remains problematic. Rather than counting devices, the Aubrey Holes more likely were offertory pits—we know that they were refilled with chalk shortly after they were dug up.

I am convinced that if Stonehenge has anything to do with lunisolar astronomy, the association between its Neolithic builders and the sky is more closely allied with theater than with exact science. Stonehenge was built to celebrate the entry of the sun god into the circular sanctuary, to chart his course as well as that of his more vicarious counterpart, the one with the silvered, oft-slivered variable countenance who migrated even farther to the northern and southern climes.

Far to the north, Callanish, on the Isle of Lewis in the Scottish Hebrides, may be even more spectacular theatrically than Stonehenge. There, when the moon migrates to its 19-year southern limit, it just barely rolls above the south horizon. It perches in a notch momentarily, then glides forward and downward on a low-angled course back into the earth. Whether this was the moment that timed an ancient ritual 4000 years ago we cannot say, though the view from the center of the enclosed ring of standing stones on the appropriate occasion is truly impressive.

Though the Paleolithic bone in France may represent the earliest dated astronomical record on earth and Stonehenge may be one of the most spectacular, there is evidence that our later ancestors applied flint to stone long before they took stylus to tablet, quill to papyrus, or brush to codex. Petroglyphs or primal notations in stone have been found all over the world. One example from this side of the Atlantic is particularly noteworthy because it comes from a time (about the second millennium B.C.) and a place (northwestern Mexico) that fits with the migration route from Asia to the Americas. Second, its content seems heavily laden with astronomy and third, these archaic markings made by semi-nomadic peoples of the desert periphery of the Great Basin of North America may offer early clues to certain concepts of timekeeping that later surface in the writing of the high cultures of Mesoamerica.

Universidad de Monterrey anthropologist William Breen Murray has spent a lifetime taking photographs and cataloguing representations of carved markings on the walls of canyons up and down the usually parched riverbeds of Mexico's northern state of Nuevo Léon. Many of these petroglyphs consist of abstract figures such as meanders (are they water or snake symbols?) and spirals (perhaps a journey, or the turning around of the year?). But some are rather more concrete; for example, the head of a reindeer with a certain number of dots positioned between the antlers probably represents the tally of a kill, while a tree in profile adjoined by strike marks might indicate the size of a harvest or gathering.

An astronomical record in stone easily reveals itself. Like Marshack's bone, if such a device were intended to count moons, then we ought to find moon-numbers, like 29 or 30 (the number of days in a month), or 12 or 13 (the number of months in a year of the seasons). This is precisely what Murray finds scratched on a number of carved bison scapulae and carved in one very special stone from the hilltop site of Presa de la Mula.

At Presa, vertical strike marks hang from six horizontal lines, each compartmentalized into vertical cells by three straight lines. They total 207, the number of days in seven lunar months. This is not an unreasonable period to record if you are a member of a society on the move, never fixed in one place for a full seasonal cycle. (At Boca de Potrerillos, another petroglyphic site 25 miles [40 km] to the east, Murray found another dot pattern with exactly the same total.)

Studied close up, the groups of markings reveal that certain number combinations repeat themselves. When added together and set off from one another

The carving in a rock face at Presa de la Mula near Monterrey, Mexico, may be an early calendar. As with the carved bone from Paleolithic France, the numerical divisions of stroke marks, which may have been etched as long as 2000 years ago by local hunter-gatherers, match the sequence of the phases of the moon.

by small circular "completion marks" they appear to record the number of days in a cycle of phases of the moon. One cluster counts from first to last quarter, tacking on the last five days until the crescent was seen to disappear. Another splits the cycle in two, measuring from first crescent to full and then from full to last crescent. Still a third array segments the phases into quarters. The way the count is presented on the stone in slightly unequal intervals leads to the conclusion that these early moon-watchers were noting the visible moon days exactly as they saw them, for indeed, when measured by true lunar visibility, some months really are longer than others. In this sense the Presa stone is more an historical record—what a modern astronomer would call raw observational data, as opposed to celestial information in its processed form, the way it would appear on our wall calendar.

Farther south and somewhat later in Mesoamerican history (about the first century B.C.) another petroglyphic form appears. In the early 1960s, at the monumental ruins of Teotihuacán in the central Mexican highlands, archaeologist René Millon of the University of Rochester excavated a petroglyph carved in the floor of an early building. Pecked into the stucco layering was a quartered double circle. Its axis aligned with a similar design carved into part of a basaltic outcrop positioned on a low hill almost two miles (3 kilometers) to the west. Measuring the alignment between the two, Millon's assistants found that the direction fit perfectly with the rectangular grid plan of the great city.

These pecked cross petroglyphs of Teotihuacán probably functioned as architect's benchmarks to orient the city so that it would lie in the correct posi-

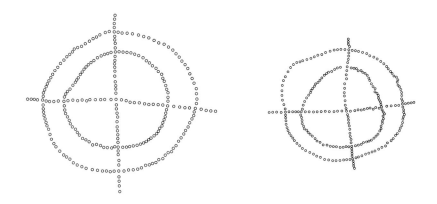

In the early 1960s, an archaeologist from the University of Rochester excavated a petroglyph carved into the floor of an early building at Teotihuacán, Mexico. The axis of this quartered double circle aligns with a similar design carved into a hill almost 2 miles (3 kilometers) to the west. These crosses are thought to be benchmarks for the architects of the ancient city. Several dozen such markers have been discovered throughout Mesoamerica. This pair is situated to mark the Tropic of Cancer.

tion with respect to the sky. But the overlapping hypothesis that these and other similar symbols at Teotihuacán and elsewhere were also used as astronomical counting devices is equally admissible. For example, though the total number of peck marks on individual petroglyphs varies, the tally on each axis almost always comes to 20, the number of fingers and toes by which pre-literate people might be expected to count. To use the language of human gesture, we can call it a "personful of days." In a few instances the total is a close approximation to 260, the so-called "count of the days" in the sacred calendar cycle, the remarkable system that served as the touchstone of Mesoamerican timekeeping.

Two more pecked crosses were discovered in a very significant geographical place in the 1970s by archaeologist J. Charles Kelley. They remain one of the most intriguing facets of the pre-literate calendar puzzle.

The Teotihuacán outpost known as Alta Vista (Chalchihuites) lies in a valley just southeast of the modern city of Durango, Mexico—on the Tropic of Cancer. Its principal temple has its corners oriented to the four cardinal points. The 28 pillars inside may have been a symbolic reference to the number of days in the visible lunar cycle. At the edge of a high plateau 6 miles (10 kilometers) to the south, two petroglyphs just like the ones at Teotihuacán are situated. Their axes are positioned to indicate the sunrise at the summer solstice. The June 21 event unfolds precisely over the most prominent peak on the eastern horizon, a mountain called Picacho Montoso. What is so important about the June solstice sun at the Tropic? If you follow its daily motion there, you will discover that at high noon the sun stands in the zenith or the point in the sky directly overhead. At that moment it casts no shadow. Then, on succeeding days, it returns back to the south at noon. Only within the limits of the two Tropics can the sun pass across the zenith to the opposite side of the sky. This principle was employed to divide the year among a number of civilizations that developed in the tropical latitudes. Outside the Tropics the sun never reaches the zenith. Thus, the Tropic itself (23½ degrees North or South latitude) can be thought of astronomically as a line of demarcation between two different modes of time reckoning.

The ability to control time would have been a major preoccupation of the Teotihuacán culture, which influenced all of Mesoamerica from the northern desert to the southern rain forest. It was this civilization that colonized Chalchihuites about A.D. 600. They must have been aware of the special way the sun behaved at that time. At least they seem to have marked it out and explicated it, in the same way that modern culture devised the system of standard time zones and later introduced Daylight Saving Time.

Today, if we stand in the ruins of Chalchihuites rather than on the southern plateau, we can still watch the sun rise over the same peak on the spring and fall equinox. The architects appear to have constructed a labyrinthine walkway that exits into a walled accessway directed precisely toward Picacho. Here, just as at Stonehenge, the ritual element may be brought into play with the astronomical. But who danced or walked there we cannot say, for all evidence about the nature of the supposed ritual has vanished along with the people. Perhaps we can imagine elaborately garbed individuals—were they the priests of a solar cult?—proceeding back and forth along the equinox path the way Stonehengers walked the accessway through Stonehenge's Heel Stone gate at sunrise on these important days of the yearly cycle. Where did they deposit their offerings? What was their route to the top of the plateau and how did the sun rites unfold 15 centuries ago? We can only speculate. All that remains for us to witness are the alignments between the architecture of ancient human beings and the sun and mountain of nature, the counts of tally marks pecked in stone and cycles of real time that measured out the planting season.

In our next chapter, when we look at evidence bearing on the ancient practice of astronomy in the Old World, facts will seem far less elusive than the proposed alignments of Stonehenge, the strike marks on the stone of Presa de la Mula, the gouged-out pits on archaic bones and stucco floors of an ancient city. When we examine Greek and Roman sundials, Egyptian star ceilings, and detailed geometrical sky models from classical Greece, we will be armed with evidence already written in texts. Then our human ancestors will begin to emerge from the dark shadows of the caves of Europe; they will solidify and take shape out of the shimmering imagery of the deserts of northern Mesoamerica. They will seem closer to us because we will stand on firmer ground when we trace our roots back to them. However, we will not lose sight of the fact that even our literate ancestors left an unwritten record of astronomical knowledge in carvings, paintings, and other iconographic forms. We have learned that when the unwritten record is all that survives, the evidence can be both suggestive and tantalizing. We can never say for sure whether such a record, so difficult to interpret, tells of archaic cultures that marked precisely what they saw in the sky. We do not know whether they used such records to predict what might happen in the future, but there is no question that the motive to record and to predict was in them, for it always makes good common sense to chart out the forces of nature witnessed in the past to be able to project into the future.

While the sophistication of today's technology is providing us with an ever more detailed picture of the universe, it also has moved us away
Greeks, Babylonians, and Egyptians used the study of the heavens to direct the business of everyday life.

3

TAPROOTS OF WESTERN ASTRONOMY

Today we are familiar with the giant telescopes and sophisticated computers with which the textbook science of modern astronomy is practiced. Supported by the laws of mathematics and physics, astronomy seeks to disclose the ultimate causes of the events we can all witness transpiring in a vast, space-bound, orbit-filled universe. But what are the origins of this scientific endeavor? From whom did we inherit our way of knowing the universe, and how were these ideas communicated to us?

from the roots of astronomy. The ancient

Most of us are aware that modern experimental science was born of the European Renaissance, but we will discover that its roots can be traced through the ancient Greeks, who gave us the gifts of logic and reason in the form of geometry—our spatial way of understanding things—to the Babylonians, whose arithmetical thinking the Greeks had inherited. In this chapter we will decipher a Babylonian astronomical cuneiform tablet so that we can explore how the Babylonians offered us yet another way to chart the sky. But our astronomy also has the flavor of Egypt, which contributed still other time-marking technologies.

We may come away surprised to learn that, for the most part, the kind of astronomy indulged in by our predecessors was nothing like the lofty discipline we read about today in popular science journals. Instead, their astronomy focused on the business of everyday life, and was directed toward the casting of omens, making prognostications about crops, fertility, events in the lives of kings and queens, wars and affairs of state. To peer at the stars through their eyes, we require a face-to-face encounter with astrology—today's impoverished, highly suspect mother of astronomy. In our Western tradition, the sky was once the home of the gods. Like colorful comic-book characters, they seem to have been capable of an extreme physical and emotional array of activities to which no mortal could possibly aspire. But celestial gods were chosen carefully to behave in correspondence with their representative stars or planets. Our ancestors' heaven was filled with the fire of the human imagination.

Long before Copernicus delivered us the sun-centered model of the universe, the ancient Greek scholars Eudoxus and Ptolemy gave us what seemed at the time more sensible earth-centered schemes. So well did their models explain what other careful observers had seen in the sky that each lasted for centuries. But these models rested upon generations of steady, long-term observational data such as we find in Hesiod's *Works and Days*.

Out of this Eastern Mediterranean world—out of 20 centuries of sharing, transmitting, and transforming ideas from the sands of the Nile and the Fertile Crescent, channeled through the intellectual conduit of the Aegean peninsula— the seeds of modern astronomical science were deeply sown. We are going to follow the growth of these seeds from germination to early maturation in the classical Greco-Roman world, where the science of astronomy first becomes recognizable to us as a set of ideas and methods whose lines of inheritance we can clearly trace. Only then does astronomy emerge as the sturdy, branching sapling that would blossom in the Renaissance—a time when ancient ideas would be integrated with a new empiricism that culminated in the now familiar schism between matter and spirit that makes experimental science possible.

Precisely what aspects of modern astronomy can we trace back to the ancient world? What have we acquired and retained from them and what have we chosen to cast out or let fall by the wayside? Given the many cultural backgrounds and borrowings that make up the civilizations that developed in the Middle East over many centuries, these are difficult and often controversial

This 15th-century colored woodcut depicts the Christian/Ptolemaic conception of the universe. It shows the earth at the center, surrounded by the heavenly realm with God and all His angels.

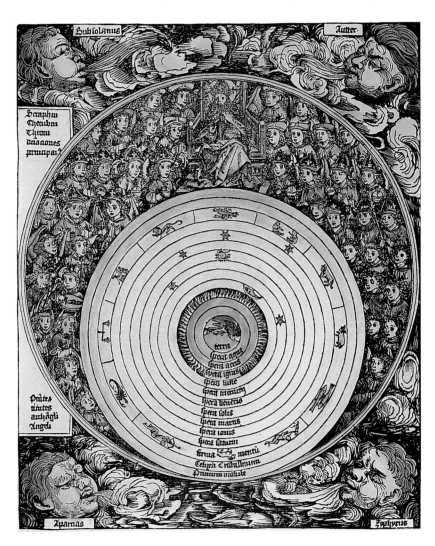

questions to entertain. Nevertheless, if we look at the record that survives the cultures that developed at this fusion point between three continents—the corners of Africa, Asia, and Europe—between the second millennium B.C. and the Christian era, we discover unmistakable signs of how modern astronomers describe, predict, and explain the motion of things in the heavens today.

Take the concept of a theory. As a way of explaining the behavior of celestial phenomena this was perhaps the greatest scientific gift of the Greeks. Their fifth-century-B.C. Socratic philosophers contemplated nature as a purely intellectual set of interrelated phenomena that operated on certain underlying principles in which they acquired an abiding faith. *Underlying* is the key word here, for the Greek thinkers of the classical period, like our contemporary scientists, believed that what we see on the surface of things is only an approximation of a deeper, hidden truth that exists in the form of unvarying principles that remain fixed in the ideal world.

What do we mean by ideal? For example, when, in the early 18th century, Isaac Newton explained that a cart rolling over a flat surface ought to maintain a constant velocity forever, never coming to rest, he was expressing the idea that in the friction-free universe that underlies the imperfect one we experience there would be nothing to interfere with a body's movement. In such a universe the natural state of things ought to be eternal motion along a straight line rather than the state of rest to which all things seem to come in the everyday world of experience. Newton reasoned that the microscopic bumps on the roadways of our world prevent us from realizing the more fundamental, simpler, underlying state of the universe. Imperfections such as friction deceive the senses by masking nature's true secrets. This is a way of viewing the natural world that Newton acquired from his Greek ancestors.

Oddly enough, the civilizations of the ancient Nile bequeathed an attitude that is quite the opposite of the Greek inheritance—a very practical means of

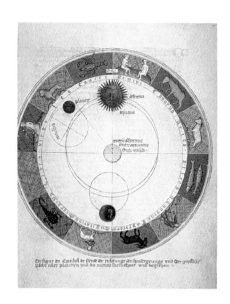

The zodiac encircles a geocentric (Ptolemaic) universe in this drawing from a 16th-century Austrian manuscript of Johannes de Sacrobosco's *De Sphaera*.

acquiring astronomical information in order to measure time. This was the annual calendar with its seasonal zodiac and its division into months, weeks (though not of the same length as our own), and the partitions of the day, all accomplished with the help of some rather clever instrumentation. If the Greek philosophers had their idea-oriented heads in and even above the clouds, the Egyptians were well grounded in making the observations that go with the theories that make up modern science.

What we had not yet acquired was the systematic, careful experimentation that has become so deeply rooted in modern science—the idea that looking for observed variations in a data set can constitute a test of a given theory. This did not happen until the Renaissance, when European scholars coupled their rediscovery of the Greek way of thought with a new empiricism they developed after centuries of questioning the Christian way of viewing everything that happened in the universe as a series of purposive acts directly perpetrated by God. Nor had we set in place beneath both fact and theory the complex mathematics to explain what we behold in the sky. Also largely undeveloped was the deeply rooted faith in science we exhibit today—that nature's truths, if not perfectly realizable in practice, can nonetheless be mathematically expressible in theory. Today the language of numbers is the tongue in which the human mind carries on its dialogue with nature. As we will see in the next chapter, this particular fragment of the mosaic of modern science never could have been set in place until all that transpired in the ancient Eastern Mediterranean world was filtered through Islam.

Strictly speaking, astronomy has no taproot—no single conduit that penetrates all the way down to an ultimate origin. Rather, as we attempt to trace scientific astronomy backward in time we will find that it divides and branches into a reticulated pattern of many converging sources. The whole story is complex, but if we would seriously go about the task of tracing astronomy's roots we must make more than brief stops in all of the diverse contributing cultures. First we need to explore the astronomical record in the cuneiform writing of the Babylonians, then the geometrical modeling of the ancient Greeks from the early Ionian school of philosophers on the north side of the Mediterranean to the later Alexandrian school on its south at the mouth of the Nile—even to the earlier celestial record-keepers farther upstream in both space and time.

Two thousand years before Caesar's reform to the timekeeping system, the Egyptians had already invented the 365-day solar year, artificial months, even hours similar to those by which we still divide day and night. Their motive for such innovations ranged from concerns as diverse and practical as agriculture to the ethereal afterlife.

Imagine living in the Nile River valley before the great dam at Aswan controlled the flow of water. A rich floodplain extended barely a mile or two on either side of the green ribbon of the Nile. Water separated desert from endless desert—an agrarian lifeline for the centuries that spanned the three millennia during which ancient Egyptian culture thrived. To early Nilotic people the

The Nile River separates one stretch of endless desert from another. To the early agrarian people living on its banks, the river's annual flooding was as important as it was predictable. It always seemed to occur when the brightest star in the sky (Sothis, or Sirius) reappeared in the eastern sky.

annual flooding must have seemed quite predictable. It always seemed to occur more or less about the time the brightest star in the sky, Sothis (Sirius to us), reappeared in the eastern sky. Then it could be glimpsed for a few moments before the light of day, having been blotted out on previous occasions for a few weeks by the solar glare. This event, the familiar heliacal rising we discussed earlier, also happened to occur about the time the sun had arrived at its greatest northerly sunrise and sunset points—the summer solstice. Each event was a convenient signal, a double time check, to reset the year clock and begin anew the count of the months. Because the months do not fit precisely into the year, some years would consist of 12 months per year, but sometimes the count could add up to 13.

Rather than live by nature, the later, more administratively minded rulers of the Nile took time into their own hands. They artificially fixed the length of every year at 365 days, dividing it into 12 uniform months of 30 days, with an extra "month" of 5 days tacked on at the end. This threw their calendar out of phase with the seasons, a fact that seems not to have bothered the Egyptians as much as it bothers us. Imagine how disturbed we would be if, devoid of a leap year cycle, one of our important holidays backed its way through the seasons at the rate of a full year every 15 centuries. If our calendar operated the way theirs did, the U.S. Fourth of July celebrations would fall in the dead of winter by the year 2750.

Each Egyptian month was divided into three 10-day weeks rather than the more naturalistic four-week months of the old calendar, in which each week corresponded to a visible quarter-phase cycle of the moon. Thus there were 36

weeks in an Egyptian year, plus five days left over. Today we might look for someone's clock on the mantelpiece, the kitchen wall, or perhaps on a person's wrist, but the first inkling that the Egyptians used a clock set by the stars to divide the night, comes—of all places—from the inside of a coffin lid. When a pharaoh died, he was believed to pass into the underworld, the same place from which the stars appear and where they disappear when they pass over the western horizon. Therefore it is logical that the star clock should appear in front of the countenance of one of such importance, for he must be carefully guided upon his eternal journey.

How did the clock work? The scheme was far different from the way our conventional timepieces operate. The mechanism is neither a set of gears moved by a swinging pendulum, nor an oscillating crystal; it is the motion of the stars themselves. For a particular week each of the 12 hours of the night is marked by the rising of a particular star or set of stars called "decans." Next week the stars change position by sliding over one hour; that is, those stars which rose to mark the first hour of the first 10-day week would mark the second hour of the second week and so on, each decan passing out of the clock at the end of 120 days, or 1/3 of a year. Thus in the star clock, which lists only 18 decans, or half a year, a particular star name in one of the 12 vertical hourly columns moves diagonally in the succeeding week.

What the Egyptians imposed upon the Nile they also did to the sky. Just as they regulated the water supply by artificial irrigation, storing the supply in man-made lakes, so too did they mark the passage of astronomical time by means that depended less and less upon the watchful eye's awareness of the vagaries of nature. Developed early in the third millennium B.C., these star clocks, relying as they did upon knowing which constellations to observe, eventually passed out of existence by the time of the New Kingdom (by the 15th century B.C.). They were replaced by water clocks—a rather significant development, for this sort of time-keeping is totally under human rather than natural control. Thereafter, in Egypt, the subject of time passed ever further out of the hands of the astronomer and into those of the architect and engineer—people more akin to their modern descendants, who built the great dam at Aswan.

THE LEGACY OF BABYLONIAN ARITHMETIC ASTRONOMY

Of all the cultures that made our astronomy what it is today, the Babylonians have left us perhaps the clearest record of their gifts—all of it in a once-alien writing form that has now been fully deciphered.

It is not surprising that the Babylonians should invent writing in clay—that they should hammer curious juxtapositions of triangular impressions we call cuneiform (cunei means wedge-shaped) onto wet, viscous earth compounds, allowing them to harden into almost eternal permanence. Clay tablets were the logical medium to carry astronomical as well as other information. Greasy clays were abundant all over the land between Tigris and Euphrates, especially after the annual flooding. As the mud dried and caked, its components created a natural medium on which to create visible symbols that could help one to remember things past. The tablets are still capable of telling us today what these people thought about more than 2000 years ago. Their laws, their stories of creation, their dynastic histories, even the passage of time are recorded in cuneiform script. Just as they were once positioned in city squares and ancient libraries, today the clay tablets reside in museum cases to be read by all the people.

According to art historian Denise Schmandt-Besserat, neither mathematician nor astronomer invented cuneiform. She believes the wise innovator was probably a farmer or trader. In bringing jars of oil to an ever-expanding market economy to trade for sacks of barley and dates, there is a problem. How to remember and record who owes what to whom? This is especially difficult if large quantities are shipped long distances. Researching the world's first bills of lading, which date from 3350 B.C. and come from excavations in Uruk and other cities in Iraq, Iran, and Syria, Schmandt-Besserat managed to crack the economic mnemonic code. These account books consisted of clay envelopes that held collections of small clay tokens of various shapes. Each configuration represented a particular kind of trade item; for example, cones represented grain, disks were animals, and a sphere with a hole in it stood for a unit of land measure. Once a transaction was agreed upon, the envelope was sealed and its contents embossed on the outside of the envelope in two-dimensional replicas of the shapes within. After generations of such use, Schmandt-Besserat theorizes that somebody conceived of the efficient idea of allowing the record on the outside of the envelope to stand alone as a statement of business transaction. In effect, that wise merchant flattened the envelope and the linear series of marks on the outside took the form of writing we see on the tablets.

The astronomical cuneiform tablet opposite is one early form of Middle Eastern writing. When 19th-century epigraphers deciphered it they discovered that this particular tablet is not about oil or grain. Instead it tells us about the kinds of astronomical records Babylonian astronomers kept and how they used their observations to make celestial predictions. A few translated excerpts read:

> If on the 6th day of Ab Venus appeared in the east, there will be rains in heaven; there will be disaster. Until the 10th of Nisan she will stand in the east; on the 11th of Nisan she will disappear, and, having remained absent 3 months in the sky, on the 11th of Tammuz Venus will shine forth in the west; hostilities will be in the land; the harvest of the land will be successful.

> If on the 7th of Ulul Venus appeared in the west, the harvest of the land will be successful; the heart of the land will be happy. Until the 11th of Ayar she will stand in the west and on the 12th of Ayar she will disappear, and, having remained absent 7 days in the sky, Venus will shine forth in the east on the 19th of Ayar; hostilities will be in the land.

> If on the 8th of Tesrit Venus appeared in the east, hostilities will be in the land; the harvest of the land will be successful. Until the 12th of Sivan she will stand in the east and on the 13th of Sivan she will disappear, and, having remained absent 3 months in the sky, Venus will shine forth in the west on the 13th of Ulul; the harvest of the land will be successful; the heart of the land will be happy.

Subjective terms like "hostilities," "happy," and "successful" would seem strange to us if we encountered them in a modern astronomy book, but do not be dissuaded by the omen-bearing qualities the Babylonians assigned to their astronomical tables. This very early text, which dates to the 17th century B.C., is nevertheless a diary of very accurate sky observations written partly in the conditional tense. Its subject is the planet Venus. Similar texts tell about the moon and planets, still others about eclipses, dates of solstices and equinoxes, and such; some offer predictive phenomena for a specific year.

The scheme for marking time as stated in the tablet opposite may seem familiar to us. In each of these statements, the planet Venus is said to disappear on a particular day of a given month and to return on another. These observations refer to the times when Venus disappears in the light of the sun as evening star and when it reappears as morning star—and the converse (the heliacal events discussed in Chapter 2). Between these stations in time the astronomers computed an interval. The Babylonians also marked stations in space (degrees of longitude and latitude) as well as time. This is indicated in other quotations in which, in a before-and-after sequence, Mars is said to stand in one position on the zodiac and then in another. The interval in between gives the degrees of separation. In still another example, astronomers record a series of lunar eclipses from which they calculate the difference between the times of occurrence of successive eclipses.

In all of these examples the Babylonian arithmetical way of forecasting is based on a very simple sequence:

PLACE + SPATIAL INTERVAL = FUTURE SPACE or
TIME + TEMPORAL INTERVAL = FUTURE TIME

The first formula charts the future place in the sky where an event ought to be observed, the second the future time when an event ought to take place. We will see this same sort of predictive device operating in the ancient Maya world. A pair of simple examples will help us understand how this form of predictive astronomy, which we have inherited in modern science actually works in practice.

The translated excerpt from a Babylonian tablet below dates from the year 3, 27. Written in sexagesimal notation this means 3 x 60 + 27 or the 207th year of the Seleucid era (104 B.C. in our calendar). The text consists of a listing by months of the year (Roman numerals, beginning with the 12th month) of the *anticipated* positions of conjunctions of the moon and the sun in the zodiac (right column). The goal is to arrive at the next month's conjunction position in the zodiac. Our document is nothing less than a new-moon prediction table. Such a table would have had practical value. We know the Babylonians, like the Greeks after them, counted days of the month from the first visible crescent in the west. Their business affairs, the setting of religious holidays, their entire schedule of livelihood depended on this count. Therefore the desire to *anticipate* when the event would take place would certainly have been present.

The Babylonians kept their astronomical records on clay tablets. This one is based on the movement of Venus—it records when Venus disappears in the light of the sun as the evening star and when it reappears as morning star. But the goal of all Babylonian astronomical predictions was astrological.

XII 29,8,39,18	2,2,6,20 Ari.
I 28,50,39,18	52,45,38 Tau.
II 28,32,39,.18	29,25,24,58 Tau.
III 28,14,39,18	27,40,4,14 Gem.
IV 28,24,40,2	26,4,44,10 Cnc.
V 28,42,40,2	24,47,24,18 Leo.
VI 29, ,40,2	23,48,4,20 Vir.
VIa 29,18,40,2	23,6,44,22 Lib.
VII 29,36,40,2	22,43,24,24, Sco.
VIII 29,54,40,2	22,38,4,26 Sgr.
IX 29,51,17,58	22,29,22,24 Cap.
X 29,33,17,58	22,2,40,22 Aqr.
XI 29,15,17,58	21,17,58,20 Psc.

I italicized the word "anticipated" above to emphasize that this is a purely predictive table. It must have been based upon years of continuous observation of which constellations of the zodiac could be sighted before sunrise and after sunset, for one can never see the sun residing in a constellation. Take for example the first place in the constellation of Aries, written 2° 2' 6" 20''' in the table, and add to it the interval on line two, 28° 50' 39" 18''':

$$2° \ 02' \ 06" \ 20'''$$
$$+28° \ 50' \ 39" \ 18'''$$
$$=30° \ 52' \ 45" \ 38'''$$

Note: the symbols ', ", and ''' stand, respectively, for the 60 divisions of the degree in the sexagesimal system that we call minutes, the 60 divisions of the minute called seconds, and the 60 divisions of the second which are no longer in common usage in our geometry.

Since each zodiacal constellation occupies 30° of the ecliptic, to get the distance into Taurus that the moon had progressed we would need to subtract this amount from the total. This would yield the number of degrees into Taurus where we would expect the conjunction to occur. This is 00°, 52', 45", 38''', which appears on the right side of line two, and so on through the remainder of the text. Next look at the intervals. Why are they all so different? The tabular model seems to be based on the fact that the sun speeds up and then slows down at a constant rate, the maximum speed occurring in month IX, the minimum in month III. Using our modern space-based or graphic way of describing this model we would approximate the sun's speed plotted against the sequence of months as a zigzag function composed of alternating sets of sloping straight lines. However, the Babylonians did it all with numbers.

Another example (in the box at right) will reveal the Babylonian power of prediction in quite a different way. It employs a simple formula based on the ability to recognize a pattern in a long series of observations. It is a transcription of an extract from a list of lunar eclipses anticipated during the reign of Artaxerxes II, a Babylonian king of the 4th century B.C. The text records the month number during which the full moon might undergo eclipse. Modern asterisks indicate that an intercalated month has been thrust into the list. This is necessary because 12 months measured by the phases of the moon fall about 11 days short of a seasonal year of 365 days. Therefore, to keep the sun and moon in step, the ancients found it necessary to regularly insert (about once every three years) an extra month into the calendrical canon.

Clearly this decadal record could only have been based on generations of observations and recordings of lunar eclipses. Only then could Babylonian astronomers have discovered the 6-5 pattern, a record broken by the fact that not all eclipses in a sequence turn out to be total. Some are penumbral, in which the moon's disk is uniformly, though more dimly, lighted by the sun. Perhaps some of these eclipses were not detectable at all and even if they were they might have been obscured by clouds or invisible because the side of the earth on which an observer was situated at the time was turned away from the sun. However, there is safety in numbers and after a long series of observations that may have extend-

Part of a Babylonian fragment recording possible lunar eclipses:

Art. 33	II		XII	40	IV
	VIII	37	VI		X
34	II		XII	41*	III
	VIII	38*	V		IX
35*	I		XI	42	III
	VII	39	V		IX
36	I		XI	43	III
	VII				IX
44*	I		XI	5	III
	VII	2	V		IX
45	I		XI	6*	II
	VII	3*	IV		VIII
	XIIa		X	7	II
46	VI	4	IV		VIII
	XII		X	8	II
Och. 1	VI			*	VII
	XII	12	IV	16	II
9	VI		X		VIII
	XII	13	IV	17*	I
10	VI		X		VII
	XII	14*	III	18	I
11*	V	1	X		VII
	XI	15	III		XIIa
			IX	19	VI
	XI	2	III	4	I
20	V		IX		VIa
	XI	Dar. 1	III		XII
21	V		IX	5	VI
	XI	2*	II		XII
Ars. 1*	IV		VIII	Alex. 1	VI
	X	3	II		XII
			VIII		
2*	IV	6	II	*	XII
	X		VIII	3	V
3	IV	7	II		XI
	X		VIII	4	V
4	IV	Phi. 1*	I		XI
	X		VII	5	V
5*	III	2	I		XI
	IX		VII		

(Note: Art. is Artaxerxes II, Alex. is Alexander the Great and Phi. is Philip.) The month number is given in the third sub-column—VIa and XIIa indicate a second (i.e. intercalated) sixth and twelfth month respectively. * indicates that there has been an intercalated month since the month given in the preceding line.) If we count the interval between Roman numerals, one of the simplest rules of eclipse prediction immediately glares out at us: once you see the moon eclipsed, do not look for another such event until six, or on rare occasions five, more months have passed.

Joseph Scaliger, born in Agen, France, in 1540, was one of the greatest scholars of his day. His major works included a treatise on the astronomy of the ancients and the invention of a system of computing time cycles known as the Julian Day scheme.

ed over centuries, the pattern would be compelled to reveal itself. Taken in this light, our simple ephemeris or timetable can be conceived as a warning table, a numerical model for making predictions that could be honed into a more definitive anticipatory device as it was enhanced by a more substantial data base.

Our modern scientific predictions work in the same way as those of the Babylonians. We devise somewhat more complex mathematical expressions that we call the laws of nature. These statements, once we insert the independent variable of time into them, generate predictions about the future. Then in a progressive way we alter and modify these equations, adding terms, inserting factors to reflect more accurate predictions that are rendered valid in the world of experience through the detection of even more subtle patterns in the observational record.

Our ancestors taught us still other ways of seeking order. We begin our day at midnight, when the hour, minute, and second hands converge in the upright position—three distinct periods measured from the same temporal marking point, which we define as the stroke of midnight. This habit of seeking the commensurate quality of time intervals—the perfect fitting together of component parts—also lies at the basis of one of our oldest calendars, the Julian Day count. Fabricated in the 16th century by French classicist Joseph Scaliger, it computes the day count from an era beginning in 4713 B.C. when the start of three ancient time cycles coincided:

(1) The period over which a given lunar phase coincides with the same date in the seasonal year (the 19-year Metonic cycle named after the fifth-century-B.C. Greek mathematician and astronomer who devised it);

(2) the period that includes all combinations of the days of the week with the first day of the year (a 28-year cycle); and

(3) the cycle of indiction, a non-astronomical cycle of 15 years' duration, relating to the collection of certain taxes imposed upon the military and instituted by the Emperor Constantine in the fourth century A.D.

Why these three unlikely cycles were chosen is not important in the present context. I cite this curious example only as a way of suggesting that discovering commensurabilities among diverse time cycles is a pervasive way of seeking order—pervasive because, in addition to the Babylonian examples, we shall rediscover that this very same principle of creating cycles of time that fit perfectly together lies at the foundation of empirical astronomy in other cultures of the world—the Greeks, the Maya, and, in Southeast Asia, the Indians.

As we have seen, our later Babylonian ancestors were particularly keen about designing arithmetical schemes to allow the planets to fit together harmoniously. How often would a meeting of planetary deities take place in a conjunction? An astronomer's diary from the sixth-fifth centuries B.C. reads:

> ... Dilbat [Venus] 8 years behind thee come back ... 4 days thou shalt subtract ... the phenomena of Zalbatanu [Mars] 47 years ... 12 days more ... shalt thou observe ... the phenomena of Kaksidi [Sirius] 27 years ... come back day for day shalt thou observe ...

Here, brought together for the first time, are the arithmetical notations we saw earlier and the habit of meshing diverse periodic cycles with one another. The words offer a formula for approximating the synodic cycles of the planets by adding or subtracting a small number of days from a full cycle. This directly observable synodic period is the interval between successive repeatable aspects of a planet with respect to the sun. For example, suppose Venus makes its heliacal rise (or first pre-dawn appearance) and is visible for only an instant near the rising sun after an interval during which it has been hidden in the glare of the sun. The planets follow more or less the same path as the sun against the starry background but with time cycles of their own. The red planet Mars, for example, takes 780 days. Every 660 days or so it slows its west-to-east advance among the stars, comes to a halt and then mysteriously reverses its direction, sometimes for several weeks, before resuming its regular course. This retrograde motion was a bone of contention in the early Greeks' attempts to devise working mechanical models of the universe that would account for it. All the other planetary tracks execute retrograde loops each in its own time period.

Venus and Mercury behave a little bit differently from the other planets. Always close to the sun, Venus never strays more than 46 degrees, Mercury never more than 26 degrees from the great solar luminary. Maya astronomers singled out this particular Venusian characteristic, and accorded Venus the distinction of being twin brother of the sun. This was probably because it usually hovered over it either as morning star before the sun rose in the east or as evening star when the sun set in the west.

Suppose we mark the date of a first appearance, say July 1, 1993. On the next day Venus will be seen for a lengthier time. As it stretches farther from the sun in the weeks following, it becomes visible for minutes, then even a few hours before dawn. Finally it stretches to its maximum distance from the sun, then slowly begins its return. A few months later it vanishes in the morning light and cannot be seen for several more weeks. Then it reappears in the evening twilight, hovering over the sunset position for but a few moments before disappearing below the western horizon as it follows the sun. In the next several months Venus repeats the maneuvers it trekked out in the morning sky. Now as evening star it stands high in the west before returning, like a yo-yo on the end of its string, back to another encounter with the bright luminary. Gone again—this time for a brief period—it finally returns back to morning heliacal rise, having completed a "Venus year." This is the synodic period of Venus, the sum of four distinct intervals, two of appearance and two of disappearance, and it totals 584 days on the average. Thus, the date of the next heliacal event would be July 1, 1993, plus 584 days or Feb. 5, 1995. Because this period fits (in the perfect ratio of 5 to 8) with the seasonal year, the fifth heliacal rise measured from July 1, 1993, would happen on (nearly) the same date eight years later, or June 29, 2001, to be exact. Using the old Babylonian calendar, which had no leap-year correction, the date would be two days earlier in the season, or June 27. Since most human activities are

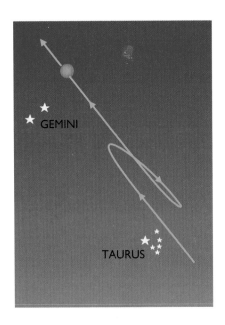

As the planet Mars passes eastward from Taurus into Gemini, it gradually slows to a halt, reverses its direction for several weeks, then resumes its normal west to east motion through the zodiac. This curious, looped path beguiled astronomers from ancient times to the Renaissance. Today we explain it by saying that the earth passes Mars as the two orbit the sun, and so Mars seems to go backward.

keyed to seasonal dates this sort of celestial harmony, envisioned perhaps through a coincidence of Venus heliacal rising events with the start of the rainy season, would serve as a logical basis to establish a calendrical reference point.

By contrast Mars' synodic period is 780 days and unlike the case of Venus (and Mercury) its synodic period, along with that of the other visible planets (Jupiter and Saturn), consists of a single lengthy appearance followed by one brief disappearance period. Once Mars makes its heliacal rise in the evening dusk it wheels away from the setting sun all the way around the night sky until it meets the sun at the opposite horizon and disappears in the pre-dawn light (except when it makes its curious backward turn called retrograde motion). Though Mars and Venus move differently they begin to repeat their moves after 780 and 584 days, respectively.

The diary passage quoted above is about how all of these observably distinct cycles of planetary movement go together in a commensurate way. And a little arithmetic, no doubt the same sort once written down on a "scratch tablet" by a Babylonian astronomer, or in a Mayan codex, will show how the passage might have been constructed. The first part of the statement is already obvious. It says that if you want to peg a Venus event in our year you need to "come back" four days every eight years. This is because a whole number of Venus years (nearly five of them) come closest to fitting our year, overshooting a whole number of years by a mere four days, as demonstrated above. For Mars, 47 of our years and "12 days more" is a somewhat less approximate formula for arriving at a whole number of Mars cycles (it turns out to be 22). The last part of the statement remains a mystery but it seems to be an attempt to tie all of these commensurabilities into the year cycle as marked by the movement of the sun among the stars. The reference point may be a pair of heliacal risings (or settings) of the brightest star in the sky, Sirius, 27 years apart.

Long interval formulae such as those in the diary would also have had a practical use. Once you had collected all the phenomena associated with a particular planet and laid them out cyclically (for example, first appearance, last disappearance, greatest angular distance from the sun), you could compute that planet's appearances for any run of years in the future by simply recopying and setting into place each of the cyclic phenomena within that period of years, adding on or subtracting the tiny aberrations, 4 days for Venus, 12 for Mars, etc. Thus you would create a powerful ephemeris to tell when and even in which constellation future events would take place; for example, the tenth heliacal rise of Venus after the hypothetical one of July 1, 1993, would happen (in Babylonian style) on June 23, 2009, and it would occur in a place backed up 16 degrees along the zodiac from where it took place originally. A more long-range set of predictions, namely when planets would arrive at conjunction with one another, could also be generated by such a table. For example, close conjunctions of Jupiter and Saturn happen about every 60 years, and furthermore they recur against the same stellar background every 960 years.

Were these abstract, esoteric-sounding numbers pure adventures of the mind, indulged in simply out of curiosity? One element lies submerged in the Babylonian calendar, and we need to discuss it in order to understand why the astronomers wrote statements like those we have been quoting and analyzing—that is the question of motive. "Thee" and "thou" are not the sorts of pronouns we use to address the planets today. We do not beckon to them to go forward or come back, nor do we believe that proper gender-indicative names like Venus and Mars signify real personae. If we overlook this deeper issue of motive and extract only the principles of commensuration and the interval-place-interval scheme as the hand-me-down mechanics of astronomy, we risk losing an appreciation of how our ancestors comprehended the firmament for themselves.

Throughout most of world history, astrology has been the generative force behind early astronomy. This was particularly true when, about 800 B.C., the Assyrian empire dominated Mesopotamia. Their dynasties were tied ideologically to a pantheon consisting in large part of spirit deities whose actions—revealed through omens—influenced the course of humans on earth, most especially that of the empire and its ruler. At least this is what the fragments of thousands of texts recovered from King Ashurbanipal's palatial library in the seventh-century-B.C. Assyrian city of Nineveh tell us. The omens are often made manifest through the action of celestial bodies. In effect the omen clauses drive the rest of the text. They are the ends for which astronomy became the means.

As much as we might laud the Babylonians for their careful observations and their astronomical predictive skills, we must never lose sight of the fact that the underlying reason for seeking intricate patterns in the heavens was to get at the future—to know in advance when an event would happen. The celestial deities wove our ancient destiny so tightly, so intimately, that we could not avoid a preoccupation with their wanderings. That the tides, the wind and the rain could be predicted by watching celestial events seems reasonable enough, but the health and wealth of kings and peasants? Hardly—at least for us. Worshippers of the heavenly abode of these deities would appeal to them by performing certain rites. The language comprising the dialogue between mortal and transcendent consisted of offerings and incantations; the implements of communication were charm and amulet rather than compass and quadrant. These people felt closely connected with what was going on around them. They experienced nature's forces directly—earthquakes and floods, miscarriages and deformities at birth, eclipses, and rainbows. They never imagined such phenomena merely as detached events in a universe devoid of meaning. All things happened for a reason—to warn them and to convey a message, either good or bad, that would guide their future. And some phenomena occurred with a more predictable regularity. For the farmer, a moonrise could tell what to anticipate in the forthcoming crop cycle. For the king, an appearance of Jupiter might signal what an encounter with the people who lived to the east would portend, those over whom he sought to extend his dominion.

The logic behind astrology is quite straightforward. A careful observer can easily become aware through everyday experience that the cycles of sun and moon are correlated with the seasons, the tides, the menstrual cycle. Then why not extend celestial destiny to encompass tides as a force of influence in the affairs of people? If we watched carefully enough, could we discover associations between the most precisely predictable occurrences on nature's stage—eclipses or heliacal risings of Venus—and the more vagarious—the outbreak of a plague or the arrival of locusts? These seem to be the sorts of questions in the minds of antiquity's courtly timekeepers, the sky specialists who composed and dictated the contents of the tablets to their scribes.

Anybody who lived and worked in ancient Babylon would have appreciated just how menacing the gods of nature could be and why their every move merited the closest attention. Before the cosmological creation myth Enuma Elish (it means "when above ...," after the first words in its text) appeared on the kind of written tablets we have been discussing, it had been part of a 1000-year-old oral tradition. The story describes a battle for control of the universe between the forces of the sky and earth deities. In ancient times, says the myth, the gods lived in intermingled harmony, just as the sweet waters of the Tigris and Euphrates rivers blended with the salt water of the Persian Gulf. Between these two watery worlds the earthly abode had been slowly built up out of the gradual silting process that still occurs in the delta. But war broke out and the gods sought to establish a new balance of power. Life in the agrarian world between the two rivers was indeed rife with tension and disorder. In winter the steady rain in the mountains of northern Iraq still produces flooding in the valleys to the south. Then the wind howls southward along the narrow fertile strip between the rivers as the drying out process begins. Just as the sun appears to finish his work he attacks the earth mercilessly, and his blazing light ceaselessly bakes the landscape into drought. Little wonder inhabitants of the fertile crescent perceived the creation of the present world as the outcome of a battle between mother earth and father sky.

A theogony is the kind of story to which child and grandparent alike could relate. It is the story of the gods' creation recounted through generational battles between parents and children out of which emerges the invincible Marduk, the hero of the gods. He slays the earth monster and from her body, slit in half and ballooned outward by the force of the wind, he creates the world essentially as we know it, with land situated below a starry domed sky floating upon a flat ocean. Marduk creates people for the express purpose of serving the gods, that they may remain contented in their present condition. The people are told to build a ziggurat dedicated to Marduk, write the tablets that tell his story, and display them in their city, thus reaffirming and disseminating the truth of the creation story. In periodic rites to the gods, Babylon's good citizens were required to reenact certain aspects of the myth over and over again so that it would never be forgotten. Marduk (Zeus to the Greeks, Jupiter to the later

Romans, Thor to the still later Nordic people) is the king of the gods and the Emperor is his earthly manifestation.

The moral of Enuma Elish seems to be that in a violent world, you succeed only by meeting force with counterforce. The way to achieve order is by violent action, the only viable means by which Marduk could bring the world about—a reasonable ideology in an urban imperial state like Babylon in the middle of the first millennium B.C. The challenge of the astrologer-astronomers, then, was to develop the skills whereby they could successfully read the omens and schedule the rites. Part of their skills depended on good record-keeping and the invention of clever predictive formulae like those we have been analyzing. And so they carefully scanned the skies above to follow their gods: "When the star of Marduk appears at the beginning of the year, in that year corn will be prosperous." This avid skywatching, turned into an astronomy based upon simple arithmetic, was passed on to the Greeks, who refined and developed it and, as we shall see, gave us much more.

THE GREEK LEGACY

We honor the Greek philosophers for teaching us how to theorize, to speculate upon our existence, to question with skepticism. How did we acquire this agenda? We need to know, for later we must ask whether only Western thinkers raised scientific questions about their own existence.

For most of the history of Greece, belief in a universe controlled by the gods was essentially the same as it was for the Babylonians from whom they borrowed. Greece too had its skilled astrologers. They are the ones responsible for the basic forms of horoscopes we still read in today's newspapers. But they also bequeathed us the logic and reason we apply in today's way of acquiring scientific knowledge about the universe. The Greeks invented a unique geometrical form of logic—a space-based way of reasoning through celestial behavioral patterns that, though developed 2500 years ago, still thoroughly imbues our way of understanding the universe. We believe in a universe of orbits and planes, of flat and warped space primarily because of our Greek heritage. They gave us, too, the concept of modeling, literally creating a mechanism with interconnected moving parts—a "simulation" that describes how natural phenomena behave. Some Greeks believed that if they looked beyond the gods the underlying principles that governed heavenly motion could be revealed. Theory, logic, models—all gifts of the Greeks indelibly written on the faces of all our modern texts about the natural world. But to understand our astronomical heritage we must open each package

To the Babylonians, Marduk was the king of the gods. According to their legends he created the world by slaying the earth monster, then he created people for the express purpose of pleasing the gods.

and examine its contents carefully to see whether the gifts have remained pristine in our culture and, if not, how each of them has been tarnished or transformed.

Take models, for instance. In early Greece they were based on everyday common sense. Thales of Miletus (624-547 B.C.) postulated that the earth is a flat disk surrounded by water. He reasoned that it must be a disk because that is what the horizon around us seems to indicate. And it must float on water because we recognize ground water surging up from below in the form of artesian wells. That there is water above us is demonstrated every time the rain falls. And as far as traders and explorers could make out, the horizontal dimensions of the earth disk were encompassed by a great ocean-sea. Such a water-bound terrestrial model is both practical and logical. It derives its explanatory power from ordinary experience.

Anaxagoras's (500-428 B.C.) world model focused more on the heavens. It consisted of a cylindrically shaped world (we live on its flat-topped surface) that floats freely in space. But the stars are appended to a sphere that rotates around the cylinder, carrying the stars below the cylinder every night, then back into view the next day. The moon shines by the light of the sun because if we watch its phases we can see for ourselves that the lighted portion of the lunar disk always faces the sun. Eclipses of the moon happen when the earth's shadow falls upon it. You can tell that because we here on earth always seem to be located between the sun and moon whenever an eclipse takes place. Whether you believe Thales or Anaxagoras, the common denominator is that simple observations of nature and common sense go together in their models.

No hidden forces, no abstract qualities—in these schemes everything makes sense to anyone who interacts with the natural environment on a daily basis. You get a picture of the world exactly as you experience it. Why? These philosophers were reared in the free-thinking, highly individualistic social environment of the Greek colonies of Asia Minor and Southern Italy. With power decentralized from the priestly hierarchy in the Pelopponesus, there was no need to dress accounts of the workings of nature in religious garb. It is easy to understand why, in the eyes of architects of modern science in the time of the European Renaissance, these Greek models of how things work seemed so impressive. If we can clearly trace one branch of the taproot of modern science back to ancient Greece, it is this habit of formulating practical, non-human-centered, mechanical models of how nature works.

Today we share with the pagan philosophers an abiding faith in the underlying principles that ultimately account for why things happen the way they do. We have rounded out Anaxagoras's cylinder and set it into motion about the sun and we have transformed Thales's static water-bound earth model into a time-dependent, precipitation-evaporation cycle. But we believe as firmly as they did that it rains when it is necessary, not because the gods will it to do so, or because the plants here on earth need water to grow. This sort of rational argumentation,

54

This reconstruction of early Athens shows the marketplace or Agora, literally "the hearth of the city." The Agora was the place where the people went to convene and commune. When the Greeks designed a city, they planned it based on their belief that it ought to be constructed from the center to the periphery. They designed their universe in the same way.

in which an idea based on unchanging underlying principles is tested in the real world, lies at the foundation of Greek science. Framing the principles in mathematical terms—thanks to the Babylonians—was another Greek gift to posterity.

If you look at the way the Greeks planned Athens, you might find further similarities with the way they believed the gods designed the universe. The Agora—literally the hearth of the city—lies at its core. There, the source of all those forces, political and economic, that drive the engine of Greek society, reside. Like the earth within the universe, the Agora is the place in society where all the people convene and commune. On the urban periphery live the different classes who compose the whole of Greek society—merchants, shopkeepers, and beyond that those ordinary citizens, the freemen who make up the Greek polis, each attended by his slaves. In the Greek concept of democracy, slaves were not regarded as people held in bondage, but rather as commodities necessary to the success of the economy. And like society, the universe was hierarchically constructed from center to periphery—orb on earth-centered orb—and among intellectuals, the physical universe was as much the subject of dialogue and speculation as the structure of the ideal city. In the Greek way of thought one has as much license to measure out the plan of the universe as the plan of the city. We shall see this theme of the city as a reflection of the cosmos resurrected in other societies as diverse as the Aztecs of Mexico and the ancient Chinese.

Geometry literally means "land measure," and it is from the practical art of surveying and city planning that the Greeks took to applying this earth-bound principle to charting out the heavens. In fact, they did it with such success that

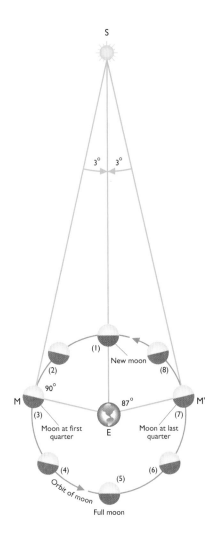

S

3° | 3°

(1)

New moon

(2)   (8)

90°

M   (3)   87°   (7)   M'

                  E

Moon at first   Moon at last
quarter   quarter

(4)   (5)   (6)

Orbit of moon

Full moon

Aristarchus's diagram demonstrating that the moon M, or M', is 1/20th the distance of the sun (S) from the earth (E).

they got carried away with the abstract demonstrative powers of mathematics. Let me illustrate with a simple celestial proof of Aristarchus (384-322 B.C.) that could fit quite comfortably in a schoolbook on geometry.

On the Greek island of Samos, Aristarchus sought to determine the relative distances of the sun and the moon. He used a diagram like the one here to arrive at his answer. He assumed that the moon (M') moves about the earth (E) in a circular orbit at a constant velocity and furthermore that the sun (S) lights the portion of the moon that faces it. The phases of the moon are disclosed to us because of our position relative to the sun and moon over the course of a month, measured in the diagram by one full circuit of the moon on its orbit. For example, when the moon is obscured in the solar glare, we cannot see it (1). At position 2 we see the thin sliver of the first waxing crescent. At 3, one half of the lighted half of the lunar sphere is turned to us. The angle EMS is 90 degrees. We call this phase first quarter. At 4, the moon is in the gibbous phase, and the angle from the sun has opened to greater than 90 degrees. When the three line up in the order moon-earth-sun we see a full moon (5) and so on back through the waning cycle: gibbous (6), last quarter (7), crescent (8), and new moon again. Now, Aristarchus reasoned that the time it takes for the moon to pass from first quarter through full to last quarter, (3) through (7), can be compared with how much time expires between last quarter through new phase to first quarter, (7) through (3). This ratio of intervals could be used to calculate the ratio ES:EM in the right-angle triangles in the diagram. For example, suppose the first interval were, say, 20 days and the second 10 days, then the larger angle MEM' would encompass 10 parts in 30 of the orbit or 120 degrees. This would make the smaller angle MEM' 60 degrees in a 30-degree-60-degree-90-degree right triangle, the sides and base of which could be tabulated. Now, Aristarchus states that observations of the moon reveal a much smaller but nonetheless observable difference between the two intervals. He tabulates the resulting angles MES and M'ES as 87 degrees, which makes the angles MSE and M'SE equal to 3 degrees. This gives a ratio ES:EM of 20.

Interestingly enough, there is no modern means by which we can detect a significant observable difference between the intervals Aristarchus claims to have detected. In fact, if we depicted the situation with a modern diagram, the sun would be placed so far off the top of the page that its rays, tangent to the top and the bottom of the lunar orbit, would be essentially parallel. Did Aristarchus fabricate the data solely for the purpose of enhancing the demonstration? Was geometrical logic being served by astronomical knowledge, rather than the other way around as we think of the relationship between the experimental world and abstract mathematics? The example of Aristarchus's proof is well worth all this elaboration because it shows that the order of priorities in scientific explanation is very different today. Like Pythagoras before him, who sought the meaning of all things in pure numbers, Aristarchus looked for reality in a geometrical diagram rather than in the sky.

56

Before we judge Aristarchus with undue harshness, we ought to realize that by concluding that the sun lay at almost 20 times the lunar distance, he was the first person to make later astronomers aware that the sun was indeed a very large body. Its disk only looks the size of the moon in the sky because it is so far away. Put at the same distance as the moon, the sun would be 20 x 20 or 400 times its cross-sectional area. Can the sun be too large to move about the world? Aristarchus must have reasoned. Might it lie at rest in the middle of the starry sky? Could the earth revolve around it instead of the other way round? This suggestion was 1700 years too early to be accepted as common sense by any rational thinker of the day. If Aristarchus had only known that his computation of the ratio ES:EM was off by a factor of 20 and that it would in fact take nearly a hundred million moons—even a million earths—to fill the volume of the sun, even he would have been amazed.

We can clearly see that when we shift focus from our Babylonian to our Greek roots we acquire one quality of modern astronomy with which we have become quite comfortable. Imagining celestial bodies as spheres moving about a void, or appended to spherical shells rotating about a common center—this was a characteristically Greek way of expressing how the universe behaved. This space-based geometrical way of perceiving the cosmos lives on today in modern maps of the solar system, the Milky Way Galaxy, and the Big Bang origin of the universe. It is the source of our concept of space travel, and it extends in the opposite direction to the microcosm of subatomic structure. As we shall see when we look at the astronomies of other ancient cultures, the framework of space is not the only ideological pyramid on which to build a universe. To reduce to a minimum the sum total of observed motions on the sky dome and to perfect a working model of these motions consisting of circles and spheres inscribed into three-dimensional space—this was the ultimate challenge of Greek astronomy from the classical period forward.

The observed celestial motions that preoccupied the Greeks were the same as those that challenged Babylonian astronomers, except that the latter, as we have seen, chose to account for them in a primarily temporal rather than a spatial mode of expression. What were these observed motions?

(1) The relatively rapid east-to-west motion of everything in the sky in 24 hours.

(2) The much slower west-to-east motion of the moon, the sun, and the planets reflected against the background of constellations, each in its own period.

(3) And finally, for the planets only, the periodic short-term reversal of this second type of motion during which the planet slows to a halt, turns westward for a time, stops again, and then resumes its normal eastward course among the stars. This so-called retrograde motion, which results in a curious loop for each planet if plotted out against the stars, seems most beguiling of all. Even by the time of the scientific Renaissance in the 16th century, it continued to defy both Copernicus and Kepler.

Claudius Ptolemy was a second-century astronomer from Alexandria who devised an earth-centered model of the universe. To explain retrograde motion, he imagined that each planet turned on an orbit the center of which moved about a fixed earth. His influence on astronomy extended well into the 17th century.

Eudoxus of Cnidus (408-355 B.C.) was one philosopher who sought to meet the challenge of reconciling these motions, a challenge which was said to have been issued by no less thoughtful an ideologue than Plato. First, to give a theoretical explanation of planetary motion, Eudoxus devised a "homocentric" model that consisted of a set of concentric crystalline spheres, one assigned to each of the seven moving bodies. Each rotated at such a rate and had its axis inclined at such an angle so as to "save the phenomena," that is, to geometrically reproduce the observed motion of each planet appended to it. The outermost spheres, to which all the others were attached, contained the stars. It turned about the earth-fixed center of the universe in 24 hours. While this model did a pretty fair job of returning all things to approximately the correct place, it failed to account for the variable rate of movement of the sun over the course of the year and, worse still, it did not explain why a planet's given retrograde loop differed ever so slightly from the previous one. Only by making his crystalline sphere model more complex—namely by adding additional spheres—could Eudoxus's followers in succeeding generations even begin to account for the subtleties of what once were perceived to be rather simple movements, now rendered more complex in the wash of fresh data gleaned from more careful observation. At one point in the modification of Eudoxus's model, Mars alone (a particularly aberrant planet) required a total of seven spheres to save its phenomena.

Ingenious as it was, the homocentric sphere model of Eudoxus literally caved in under the sheer weight of complexity. For the intellectual substrate of which Eudoxus was a part, no gods, whether they meddled directly or indirectly in human affairs, could be imagined to create a universe so complicated as the one the Greek philosophers had tried to idealize in this sort of model.

Even more imaginative was the model that came out of Alexandrian Greece in the second century A.D. This idea, devised by Claudius Ptolemy, retained the earth at the center of a spherical universe but handled retrograde motion in a more elegant way by imagining each planet to orbit on an epicycle, the center of which moved on a bigger orbit, called a deferent. Every time a planet, say Mars, reached the inner part of its epicycle and aligned with the earth, it appeared to execute a brief backward loop—retrograde motion. This effect can be compared to watching a piece of chewing gum stuck to a moving bicycle wheel. As the rider passes by, the viewer fixes his or her eyes on the path of the gum. When it comes in contact with the ground and the wheel rolls over it, the gum momentarily reverses its direction, then pitches forward again and passes over the top of the wheel.

Lest we become lost in the ethereal Aegean world of model-making, we need once again to remind ourselves that the ends served by saving the appearances were primarily those of astrology, for the state at large demanded the omens and predictions on which they depended. Indeed, Ptolemy is as well known for his astrological writings as he is for his astronomical prediction tables. What we

Nicolaus Copernicus, born in Poland in 1473, started out as a canon in the Church. There he was exposed to the ideas of the Renaissance that subsequently led to his interest in astronomy. He believed in the Aristarchean theory that the earth moved around the sun, and he worked hard at furthering this belief through observation and calculations. When his own works were finally published, they were immediately banned by the Church for contradicting the time-honored belief that the earth was the center of the universe. This ban was not lifted for nearly 200 years.

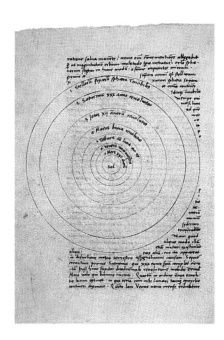

This page from the original manuscript of Copernicus's *De Revolutionibus Orbium Coelestium* illustrates his theory of a sun-centered universe.

also ought to appreciate about these speculative models of our philosophical predecessors is the way they consistently clung to the belief that circular orbits could be re-envisioned in a spatial framework to give an account of the behavior of things. It would be inappropriate to label Ptolemy's geocentric universe idea as wrongheaded because the earth really isn't the center of everything, or because objects can't travel on epicycles at the center of which there is no material body to attract and hold them in place. We must resist the present-centered habit of transplanting our notions and ideas across the vast sea of time and implanting them in the heads of our forebears. For example, the concept of gravitation, and the idea that forces can act at a distance along lines interconnecting celestial bodies, was not imagined until the time of Newton, who lived 15 centuries after Ptolemy. Never mind that the epicycles were not accurate enough to account for small-scale motions. Like that of Eudoxus, Ptolemy's model was improvised by putting the earth slightly off-center and later by adding epicycles to epicycles.

The Ptolemaic geocentric cosmic model withstood the test of observation well into the scientific Renaissance, when Copernicus replaced it with the heliocentric (sun-centered) theory of the universe and Kepler's analysis of more precise data revealed that planetary orbits were ellipses, not circles. Greek scientific skills were grounded in rationalism as well as the ability to modify a theory in the face of new information. They were erecting the universe they desired, one controlled by hidden principles in the hands of gods, who had far more important celestial business to tend to than to meddle in human affairs.

The ancient Greek models of the universe that we have been discussing are as mechanical and machine-like as the ones our scientists devise today. We say the brain is like a sponge, the nervous system like a computer, and that the atom operates like its macroscopic counterpart, the solar system, as a kind of celestial billiard game. In his work *Gears from the Greeks*, science historian Derek de Solla Price speaks of Greek models of the universe as *simulacra*, literally semblances of reality. Such shadowy representations take on concrete dimensions in a most vivid form in the Antikythera mechanism. This heavily corroded bronze artifact was found in pieces off the coast of the island of the same name by divers at the turn of the century. Unlocked from its marine encrustment, the pieces were put back together to reveal an assembly of toothed gears. Price's analysis consisted of counting the gear teeth and attempting graphically to reassemble the wheels. He concluded that the device, dated to A.D. 87, was a working machine model that not only calculated but also mimicked the Metonic cycle as a physical theory. One of the many commensurate time cycles undiscovered by the Babylonians, this cycle returned a visible full moon to the same date of the seasonal year. In the year 432 B.C. the Athenian astronomer Meton was the first person to point this out. Thus Price found that 19 rotations of the main wheel on the Antikythera mechanism coincided precisely with 235 rotations on another, just as 19 tropical years fit precisely into 235 lunar synodic months. The set of nearly a dozen bronze wheels is really far more complicated than I have made it out to be, and what it actually reveals about Greek notions of time is not completely clear. Nonetheless, there is a ring of Babylonian familiarity in the way it seems to fit celestial cycles together so neatly.

The Antikythera mechanism is important because it narrows the gap between ancient Greek and modern astronomy. Prior to its discovery no one would have anticipated gears from the Greeks, much less gear-like mechanisms of how the universe operates—at least not before the Renaissance. The Tower of the Winds in Athens is another concrete simulacrum of nature (or more correctly a collection of simulacra). This structure, centrally located near the Roman Agora of

Early in the 20th century, a diver in the Aegean Sea brought up a mysterious mechanism from a wreck near the island of Antikythera. The shipwreck was dated to approximately A.D. 87. When the device was examined, it was possible to see an inner assembly of toothed gears that seemed related to astronomical data. This artifact was the first indication that the Greeks used a mechanical instrument to express how the universe worked.

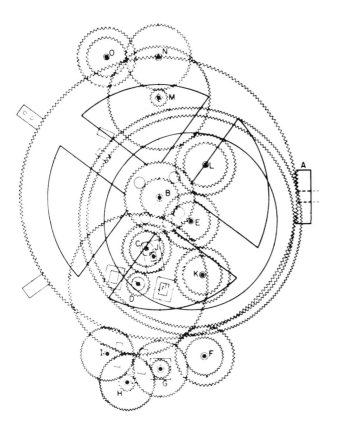

The Tower of the Winds, the shell of which still stands in ancient Athens, once housed a water clock made up of tanks with water dripping through at a measured rate. It also bore an elegant and elaborate sundial on the exterior. Still visible are the series of carved figures that represented the prevailing winds and a bronze weathervane on the top of the tower in the form of a Triton whose wand would point toward one of the carvings.

Athens, seems to have functioned something like a modern hands-on museum of science. It once housed a water clock that consisted of a flotation tank with a hole through which water escaped. Another tank fed the main unit via a slow drip. A float in the lower tank connected to a series of weights and pulleys which drove the clock. Sundials on the outside of the tower echo an earlier era when changes in the day and year were charted out by the moving shadow cast by a gnomon, or vertical rod. The shadow's path served as the analog device that duplicated daily as well as annual solar movement. Though no part of the clock survives, a Roman traveler has given us a good idea of what its star-studded face must have looked like. He tells us that it was decorated with 366 lapidary talismans and inlaid with blue stones from Yemen. What an impressive sight it must have been! Like an orrery or modern planetarium, all parts moved around in pristine perfection, just the way the Greeks imagined their real counterparts do in the sky. From the beauty of its design, just as in the planispheric astrolabes of Islam that we will encounter in the next chapter, one senses that making the working simulacrum a thing of beauty was as important to the designer as its power to explore how the universe works. And as we shall also see, the development of sophisticated astronomical instrumentation was but one contribution that emanated from the new religion that swept over the Mediterranean world like an engulfing tide between the seventh and 14th centuries.

Using a quadrant, an Islamic astronomer observes a meteor passing through the night sky above the minarets of his city. The astronomers
the spherical astrolabe, the armillary sphere, and the constellation patterns and star names we still use today.

ISLAM: SURVEYOR AND PURVEYOR

"Lo! in the creation of the heavens and the earth, and the difference of night and day ... are signs (of Allah's sovereignty) for people who have sense." (Koran II: 164)

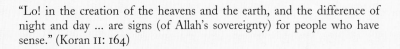

Albategnius (al-Battânî), the Arabian astronomer of the late ninth century A.D., once said that the science of the stars came immediately after religion because it alone recognized the oneness of Allah and the highest divine wisdom.

The Islamic empire, motivated above all by a deep and steadfast religious tradition, at its height stretched from Spain to the borders of India. But contrary to what many historians say, the world of Islam did not simply preserve

of the ancient Islamic empire gave the world

astronomy in the period between the decline of the classical world and the Renaissance, simply waiting to hand it over to Western Europe at the close of the Dark Ages. Rather, Arab astronomers received the gift of geometrical astronomy from the Greeks, cultivated and developed it, and integrated it along with Persian and Indian astronomical contributions into their own philosophy. Their Allah was the begetter of a more personal, human-centered universe than the one they inherited from the classical world. Islam gave back to modern science both algebra and the astrolabe. And most constellation patterns and star names that we recognize today are old Arabic inventions.

Religion motivated both Islamic architecture and its calendar in esoteric as well as highly utilitarian ways. Just as the Egyptians aligned their pyramids astronomically, Islam developed the *q'ibla*, a precise method for celestially orienting all sacred sanctuaries to Mecca. Islamic astronomical tables charted the first lunar crescent so that Muslim astronomers could keep track of the appropriate time to worship by the moon rather than via our more familiar solar-based calendar. Their books, tables, and instruments are as much great works of art as functional artifacts.

An intimate knowledge of the outdoors both above and below came quite naturally to the purveyors of this new astronomy. Before they came out of the Arabian desert in the late seventh century, unified by the prophet Mohammed, the Muslims were only a collection of tribes related by family ties. But all had spent centuries navigating by sun, wind, and stars across the vast expanse of desert.

A natural interest in numbers, combined with the need to regulate a life of daily worship, accounts in large measure for the careful corrections Arabian astronomers made to previous observations. For example, the star catalog of Ulugh Beg, the Mongol grandson of 14th-century Samarkand conqueror Tamerlane, contained well over 1000 entries. His book culminated several centuries of careful watching of star positions. His observatory was a scientific installation of truly grand proportions, a building 100 feet (32-meters) high, housing a 66-foot- (20-meter-) long shaft that calibrated the sun's noontime altitude with remarkable accuracy.

At the other end of the Islamic sphere of influence, a Moorish king of Castile, Alfonso X (called "the Wise"), who gave up his crown to turn to sky and instrument, had constructed his own set of tables three centuries earlier. Alfonso once said that had he been present at the Creation he would have made some useful suggestions to Allah about how to order his work! His careful tabulations led to the determination of the length of the year, the size of the earth, the change of position of the solar apogee (the point most distant from earth), the shape of the moon's orbit, the inclination of the ecliptic to the equator, and the period of precession of the equinoxes. Calculations of all of these earmarks of our spatial universe were thus vastly improved. Without such complexities and details, Renaissance Europe never could have developed the sun-centered theory of the solar system.

The first serious collections of astronomical instruments that combined the functions of both the lensless telescope (true telescopes that magnified the image were not invented until the early 17th century) and a rudimentary form of computer were developed by Islamic astronomers. Designed to acquire precise data about the motion and position of celestial objects, these devices foreshadowed the culmination of our later technological penchant to build bigger and better pieces of scientific equipment.

Today we draw a clear distinction between an observatory and a planetarium. We think of an observatory as a building that houses an instrument for following and amplifying energy impulses (either by direct observation or more usually with a detection device) that come to us from objects all over the universe. On the other hand, a planetarium is both a theater and an analog computer. We use it to duplicate what we would see in the night sky at a given time and place on earth. Many of the first astronomers' instruments closely combine the function of both observatory and planetarium essentially by simulating sky activity on either a flat or curved surface, as well as employing the actual sighting of a sky object to set up the simulator. A sundial, either of the type used by the Egyptians or the Romans, is the simplest form of a hybrid observatory-computer. The shadow tip is the analog of the sun that moves on an inverted bowl-shaped sky surface, all marked out in appropriate coordinates. But the instruments designed in the Arab world during the European Dark Ages were far more complex and much more ornate in design. And they provided far more precise celestial information to the penetrating eye than their classical predecessors.

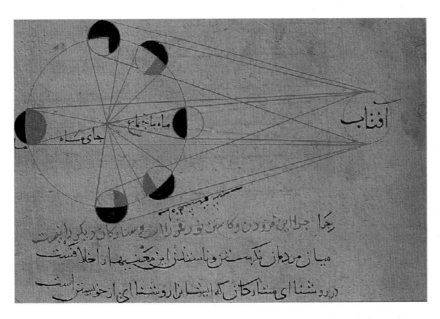

This 11th-century diagram by al-Bīrunī, one of the greatest Islamic astronomers, illustrates eclipses of the moon.

One look at the busy astronomers of Istanbul on page 66 is enough to convince us of Islam's preoccupation with precise technology. What were they looking at, and what knowledge of the sky were they able to acquire from their observations? And how did they use this knowledge? To answer these questions we need to dissect a couple of the tools pictured in this colorful work.

The astrolabe (it means "star-taker") is one of a number of astronomical measuring instruments developed in Islam that lay at the source of the precise data that appeared in collections of tables all across the southern Mediterranean world from the 10th to the 15th centuries. Within a century, the

An Ottoman manuscript from the late 16th century shows astronomers busy using a variety of instruments in their Istanbul observatory.

Mongol advance from the northeast would reduce Muslim science to smoldering ashes for Renaissance European scholars to pick over.

Invented by the Greeks, the astrolabe came in many forms. Combining the sighting properties of a telescope and the figuring capacity of a computer, this machine basically was used to tell time. The viewer looked by day or by night through a pair of sighting holes at opposite ends of a rod, mounted on a circular plate that looks like an overgrown pocketwatch hand held by a long chain. Actual dimensions ranged from a few inches up to a foot and a half. The face of the watch was outfitted with a series of star-map plates that could be removed and substituted like compact disks, one for each appropriate latitude where the user might journey. The disks themselves consisted of a flat stereographic projection of the sky onto the celestial equator (the extension of the earth's geographic equator onto the sky). The central hole marked the fixed position on the celestial sphere approximated by the Pole Star. On top of this, another plate gave basic coordinates in horizon, equatorial and ecliptic systems. Positions could be read off the rotatable ruler once the appropriate object was sighted through the peepholes. Pointers direct the viewer's eye to the brightest stars. Another circle on the flip side of the star clock served to fix the hour of the night, the day of the month, and the position of the object sighted on the zodiac; the latter function offers a tentative clue to one of the more mundane uses of such an instrument—astrological forecasting!

Of all the parts of an astrolabe, the finger-like pointers, which appear variously as dragons' teeth or serpents' tongues, seem especially foreign to those of us unaccustomed to mechanisms whose mechanical reference frames are fashioned out of animal body parts. Science historian Owen Gingerich has rooted out astrolabes with birds, bulls, horses, whales, and bears gesturing and posturing on their clock faces. In one example, warriors dance about as the tips of their swords do the pointing, while another (in the National Museum of American History in Washington) utilizes dog heads, the protruding tips of their panting tongues excitedly showing the way of the stars. Appropriately enough, Sirius, the Dog Star, brightest in the sky, is a major feature on this particular instrument.

Anyone who has visited a science or historical museum is struck by the ornate nature of ancient measuring instruments. How different the aesthetics expressed in these old tools of measurement seem compared to our own. Where they crafted brass or gilt workings inscribed with fanciful filigree, we substitute transparent plastic overlays with only the most necessary and efficient markings needed to operate the equipment. Such revelations not only tell us a great deal about the Islamic conception of beauty, but also emphasize the relationship between science and artistic endeavor that is possible among cultures of the world.

Another Islamic-developed instrument from which we derive our modern library sky globe of comparable size, along with our much larger, more modern planetarium theater, is the spherical astrolabe and its close relative, the armillary sphere. They, too, are as much works of art as artifacts of science. The latter is so named for the many concentric armillae—literally bracelets—or rings that comprise it. Again, in such an instrument, one sighted through opposing pairs of peepholes on opposite sides of the innermost ring, which could be rotated and tilted with respect to an outer set of rings. Graduated marks on the rings yielded time and coordinate positions. To avoid direct solar glare, shad-

The astrolabe was originally invented by the Greeks as a means of telling time. It consisted of a series of star-map plates that could be removed and replaced according to the user's location. The central hole marked the fixed position of the Pole Star. Pointers (often artistically designed) direct the viewer to the brightest stars—in this 15th-century example, it is the tongues of panting dogs that do the pointing.

ows cast by the higher half of the observing ring upon the lower could be used to mark the date of the equinox.

Other machines in the Istanbul observatory painting include the quadrant, the sextant, and the octant for measuring altitude, along with an early form of the surveyor's transit, or theodolite, that our highway road gangs still use for surveying today.

Just as we are indebted to them for giving us a numbering system that included the zero (the Maya had it, too), we also owe practically all the names of our constellations and stars to the Arabs. Aldebaran, Alpheratz, Algol—so many stellar designations on modern star maps begin with the Arabic article "the." Algol, for example, means "the ghoulish one," because it winks on and off every three days, a fact attributed by modern astronomers to the eclipse of a brighter by a darker star.

These star names tell virtually everything we need to know about the constellation in which they are located. Take Orion, for example. Our Hunter originally was termed al-Jabbār, the Giant—our word for algebra, the idea of unifying broken parts into an enlarged whole, has the same root. Bright red Betelgeuse, "Ibt al-Jauzah," is the "armpit of the central one" and is also called the shoulder, arm, or right hand, of the giant. Rijl Jauzah al-Yusrá, which survives today as Rigel, is his left leg, an even brighter star colored blue. The closely gathered line of three bluish stars that form Orion's belt were together regarded as the golden nuggets that lay at the middle of the constellation. Each had its own designation. Mintaka, on the right, means "belt," while Alnilam in the middle is the string of pearls set at the center of the belt. Last to rise, Alnitak is the girdle. Up in the other shoulder lies Bellatrix, the only prominent star in Orion that does not commonly bear an Arab designation. But on old maps it is al-Murzim or Mirzam, the "Roaring Conqueror." Less luminous Saiph indicates Orion's fainter leg. Oddly enough, Saiph means "sword" even though it is quite remote from the place where we might locate the giant's weapon today. Remember however, Arabs used long swords. The brightest star in the faint and fuzzy sword (which houses Orion's great nebula) is Nāiral Saif, literally translated as "brightest one of the sword." Al-Maisān (today Meissa), apparently the result of an erroneous juxtaposition with a star in neighboring Gemini, once was the Head of the Jauzah, or Rās al-Jauzah. Finally, the tidy little string of stars above Orion's right shoulder that represents the lion's skin the Hunter holds

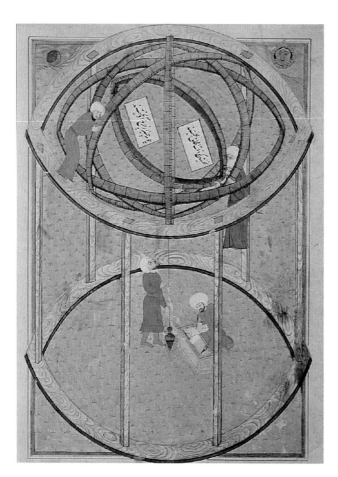

A late 16th-century illuminated manuscript shows Arabic astronomers within a giant armillary sphere.

up, were collectively called al-Kumm (the sleeve). Curiously, in Amazonian and Caribbean mythologies we will discover that Orion also emerges as a character of gigantic proportions, except that in some cases he is one-legged.

In like manner, most of the stars surrounding Orion derive from the same old Arabic nomenclature. al-Shi'rā, the Arabic counterpart of the dog-headed Egyptian deity Anubis, became our Sirius, while Aldebaran means "the bright one of the follower," presumably a reference to the Pleiades, which rise in the east moments before.

Why put a giant in the sky, or a scorpion and a bull, for that matter? There are several reasons. First, imagined star patterns were part of a huge, self-motivating text that kept local mythologies alive. The vivid imagery sketched out by connecting star to star served as the ideal illustrated text to enhance stories about real life (like Hesiod's *Works and Days*) told by sheepherders to their own children gathered about the fire under the dark sky. But there was also a more practical reason. Nomadic people like the ancient Arabs could navigate their way from oasis to oasis across a desert as barren of guideposts as the middle of the ocean by noting the points where stars rose and set and where they came to the zenith. (The Polynesians were not so different, except that they journeyed over water rather than sand.) Early Arab travelers marked their latitude by observing the Pole Star with the aid of a string knotted in several places and held vertically in line with Polaris by a fully extended arm. From the stars' changing positions at different times of the night, you could anticipate when the rains might come to fill the watering places and when your animals would become fertile.

Practically all the names of our constellations have Arabic origins—Orion, for example, was originally termed al-Jabbār, the Giant; the clearest star in his sword, Nāiral Saif, literally means "brightest one of the sword."

Finally, the wheel of the constellations, the lunar mansions and their attendant planets, also functioned to predict good times and bad times. To maintain a normal life, many of us swallow vitamins or pills to control our blood pressure or cholesterol. Our predecessors garnered times and positions with astrolabe and armillary sphere. Thus they tried to attain an even keel with the cosmos by anticipating the will of the gods and adjusting their behavior accordingly.

Many pious Muslims, including Mohammed himself, had suppressed astrology just as we have turned our heads away from the historical fact that divination by the stars was almost always an integral part of early scientific astronomy. We acknowledge the archaic practice of folk astronomy, but beyond admitting that people gave mythical names and attached marvelous stories to every object they followed across the sky, why are we so wary of exploring the metaphysical foundations of our contemporary cosmic beliefs?

69

Being both deterministic and monotheistic, like the early Christians, Muslims tended to reject some of the fruits of early astronomy acquired from the Greeks—the casting of horoscopes, for example—which accompanied the scientific packaging of geometry and rudimentary instrumentation. Nonetheless, some of the very greatest Islamic astronomers, among them al-Bīrunī (A.D. 973-1050), did in fact practice the occult art. In his *The Book of Instruction in the Elements of the Art of Astrology*, for example, al-Bīrunī outlines the possibility of predicting meteorological events—floods, for example, and earthquakes—along with the behavior of plants, animals, even people. This is all reasonable enough given what we have said about the function of preliterate astronomy. Later, al-Bīrunī goes on to admit that some predictions have origins that can never be known. When an astrologer passes this metaphysical boundary, al-Bīrunī tells us, then he is on one side and the sorcerer is on the other. At this point the omens have nothing to do with scientific astrology even though the stars can be referred to them. Operating very much in the medieval tradition, al-Bīrunī seems to have believed in genuinely detectable powers of astral influence, such as celestial rays that emanated from each star or planet, causing us here below to vibrate sympathetically or antipathetically with the cosmos. He was a true stellar determinist.

So many of the world's first astronomies seemed preoccupied with the idea of aligning the environment they built with the sacred one they worshipped in the land and skyscape. The Koran says: "... to Allah belongs the east and west. He guides whomsoever he pleaseth into the right path."

The notion that the world is divided into sections gathered about a sacred center was developed by the ancient Babylonians, whose gods were assigned patronage of different territories surrounding the ancient capital. The Etruscans enhanced the scheme into a 16-fold division of the universe, with special temples designed to face specific deities. Far away, in Peru, the Inca capital of Tahuantinsuyu (Cuzco) literally means "the four directions of the universe," and in Mexico, Tenochtitlán was regarded by the Aztecs as the world axis whence the gods derived their sustenance from the sacrificial blood of its tributary populace. The ancient tradition of the Islamic Middle East held that the holy mosque was the center of the circle of which everything outside was the inhabited world. In Muslim thinking, the four directions were halved to make eight sectors of the world. This gradually proliferated to 11-, 12-, and 13-sector schemes when it was decided, for example, that two important cities like Medina and Damascus could not occupy the same region.

Anyone with the slightest knowledge of Islam is aware of the worshipper's obligation to face Mecca, the holiest place in the world. But which way is Mecca? Here is where Muslim astronomers played an important role. They were assigned the task of determining the *q'ibla*, or the direction to the holiest of places where Mohammed established a new religion after the *hegira*, or flight from Mecca in A.D. 622. The Old Testament says that prayer was offered thrice

Moslems gather round and pray at the sacred shrine, or *K'aba*. By Islamic tradition, their shrines are positioned in perfect alignment with the will of Allah, a method known as *q'ibla*. Originally, the determination of q'ibla was made by sighting celestial bodies.

daily in the direction of Jerusalem and from this ancient Jewish tradition Mohammed likely altered the q'ibla. In the Book of Daniel (6:11) we read: "When Daniel knew that the document had been signed, he went to his house where he had windows in his upper chamber open toward Jerusalem; and he got down on his knees three times a day and prayed and gave thanks before his God, as he had done previously." The unpopularity of his early teachings among the Jews may have caused Mohammed to choose Mecca as the new holy site.

Given the veneration they held for mathematical precision, we can appreciate the trouble Islamic astronomers must have taken to set the *K'aba* or shrine in perfect alignment with the will of Allah. Only by watching the stars in the night sky and the daytime sun could they indicate the true path. The K'aba in Mecca itself has its east and west facades aligned to the sunrise at the summer and sunset at the winter solstice. Its south face is directed to the rising of the bright star Canopus, a fact specifically mentioned in medieval texts. The directions of the four winds of the classical world also fall along these alignments in Morocco, where the sacred direction is close to true east. There, the sunrise at the equinox was employed to line up the mosques, while in parts of Iran and Asia the equinox sunset, which lies true west, served the same purpose. Even the extreme rising and setting points of the moon may have been involved.

We know little about how the actual determination of a q'ibla proceeded, but in the earliest period it would have had to be acquired by sighting celestial bodies that lay in the general direction taken by a pilgrim walking to Mecca. These celestial objects may have been venerated themselves—like the star of Bethlehem, for example; but by the ninth century the whole matter of direction-finding seems to have been handled mathematically and with a minimum of skywatching. Only the direction of the local meridian needed to be determined. Islamic historian David King has analyzed a q'ibla table from Damascus that reads the angle between the meridian taken to the south and the arc of a great circle to Mecca as seen from any locale of known latitude and longitude. To solve the problem one takes altitudes of the sun with astrolabe or quadrant. Because this all happened before the development of mechanical clocks (in the 15th century), the results were not terribly accurate. Longitude determination requires that events be precisely timed, for example, by observing a total eclipse simultaneously from two localities a known distance apart.

The reliance on new quantitative astronomical methods, which were developed after the Islamic organization and assimilation of classical knowledge, was not at first accepted by the old priests, one of whom commented that astronomers now took their knowledge from infidels like Euclid and Aristotle. The older, more earthy (and substantially less accurate) folk practice of acquiring the q'ibla consisted of instructions such as: Stand with the stars of the Plough behind your right ear when the lunar mansion (Han'a) is directly back of you, the Pole Star on your right shoulder, the East wind blowing at your left shoulder and the West wind at your right cheek. This gives us the impression that prior to the Islamic improvements on Greek and Persian science before the 10th century, most ordinary people were rather well in tune with the natural environment.

The Muslim calendar also began with the hegira of Mohammed which has been calculated by Muslim scholars to have happened on July 15, 622. However, time in the Islamic sense always was, and still is, to a large degree, regulated by the moon. Like the Christian year, the Islamic one consists of 12 months; but, Muslim timekeeping was more resistant to the compromises of bureaucratization. Months were not begun and ended by counting days on an officially codified calendar page. Instead, like the ancient Greeks and the Ngas of Africa, those in charge needed to make an actual sighting of the first crescent moon in the western evening twilight. Imagine calling your local observatory to find out when to expect your paycheck or when your first-of-the-month car payment actually is due. Moreover, not every community need begin the month on the same day. And because a lunar year of 12 months is about 11 days short of a seasonal year, it also means that New Year's Day backs up in the seasonal cycle a full turn every 32.5 years. All of these time warps do not seem to have greatly perturbed anyone. Once again, adherence to the fixity of time in the seasonal year emerges almost exclusively as a latter-day Western preoccupation. We will discover that this problem seems not to have concerned Maya or Aztec timekeepers either.

The modern, more formal version of the calendar (adopted in the early Roman Republic) also has a familiar Mayan ring to it. It opts for alternating 29- and 30-day months, the last month containing 29 or 30 days according to a 30-year formula in which a majority of years (19) consist of 355 days and a minority (11) of 354. This recipe sets 360 synodic months equal to 10,631 days, one month thereby averaging 29.530555 days. Because the month once began with a lunar sighting at sunset, this also became the logical time to start every day. Again the old tradition of following sky deities was used to map out sacred time to tell the faithful when to pray, as well as sacred space—to tell them in which direction to offer their prayers. From the ancient Jewish custom of praying thrice daily, Islam developed more elaborate tables and instructions to indicate the appropriate hours of prayer. Initially this was all determined by sunwatching. The hour of the first of five daily prayers begins at sunset, the others coming at nightfall, daybreak (completed before sunrise), just after midday and, oddly enough, when the shadow of any object is longer than it was at noon by the length of that object. The last prayer period usually must be concluded when the shadow has increased another object length.

My digital timepiece is a long way from shadow-watching. The notion of determining the time to pray by watching the sun in the sky has today faded to insignificance. As Christianity once signaled the hour by chimes and church bells, Islam does so largely by recorded chants from the minaret. Above all, we must not forget that all across Europe and the Middle East, the demands of religion rather than science ultimately led to mechanical, precision timekeeping.

One Islamic measure of time worth mentioning which has not survived in our calendar consists of the lunar mansions. These were 28 star groups stretched out along the lunar orbit rather like the 12 that make up our zodiac, but some are similar to the Chinese system. Derived from India, these marked the position of the moon on successive nights of visibility throughout the month and they played a profound role in making astrological predictions.

Despite their decorative instruments and strange-sounding star names, there is much that is familiar in the substance of the astronomy that Islam bequeathed the world during the millennium when its star shone brighter than all others in the Eurasian hemisphere. In the era during which they thrived, Islamic astronomers compiled more catalogs of celestial observations than all the civilizations on earth that preceded them combined. They also found new ways to apply mathematical and computational methods to sky motions. To achieve such accuracy they advanced the scientific measuring instruments of the classical world to unprecedented precision. This led not only to more accurate systems of timekeeping, but also to the first serious criticisms of the thousand-year-old Greek models of the world, like Ptolemy's earth-centered theory, which, because they could no longer save the phenomena, would ultimately disintegrate during the Renaissance, to be replaced by Copernicus's sun-centered model of the solar system.

But where the Greeks saw their models as theoretical abstractions designed simply to predict where and when events in the sky would take place in the future, the Muslims believed theirs to be physically real phenomena that existed in a three-dimensional, space-based domain, every part of which the true believer who "submitted" (as the word Islam itself implies) to the will of Allah could experience.

An 18th-century Chinese geomancer's compass, used in divination, has a magnetic compass in the center and the eight trigrams of Taoist tradition in rectly, the user reads off the positions in the various rings along a radial line. Some of the circles refer to celestial objects and astronomical cycles.

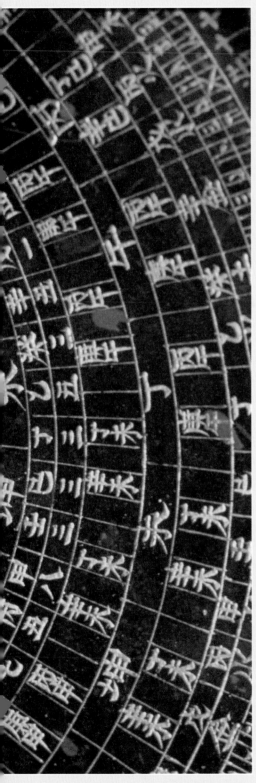

5

ANCIENT ASIA'S STELLAR BUREAUCRACY

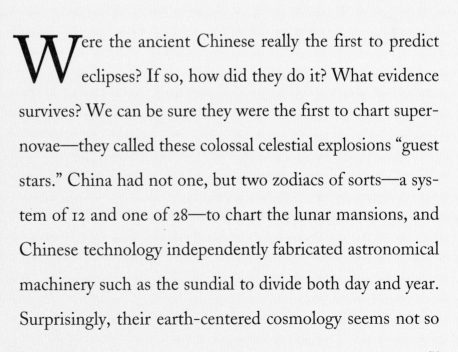

Were the ancient Chinese really the first to predict eclipses? If so, how did they do it? What evidence survives? We can be sure they were the first to chart supernovae—they called these colossal celestial explosions "guest stars." China had not one, but two zodiacs of sorts—a system of 12 and one of 28—to chart the lunar mansions, and Chinese technology independently fabricated astronomical machinery such as the sundial to divide both day and year. Surprisingly, their earth-centered cosmology seems not so

the innermost ring. When the compass is set cor-

75

very different from that of our old friend Eudoxus, the fourth-century B.C. Greek theoretician.

Despite these parallels with the West, we know that conservative, inwardly directed Chinese astronomy—unlike that of Islam—lay well beyond contact with the rest of the world, and it differed from the Western tradition by being a government activity. If the focus in China was bureaucratic—keeping track of the emperor's stars to decide how he should behave—India's first astronomies and cosmologies were geared to a more abstract religious focus. Time in the living universe was considered to be but a breathspan in the life of the creator-god, Brahma. The world ages he aspired were marked by great planetary conjunctions and royal astronomers needed to be able to forecast them with high degrees of accuracy, for life depended on it. At the head of the planetary list was Jupiter, whose regular meeting—via planetary conjunctions—with the other wandering lights in the sky served to set the hands of creation's clock.

Throughout Asia we will find that the time organized according to the stars ran in cycles rather than on a straight line, the way we tend to think of it. Evidence for an Asian interest in the sky goes back nearly 5000 years. Star-laden texts are written in many forms, on wine jugs, tortoise shell, and silk. The earliest records dug out of archaeological sites in Qinghai province consist of ceramic fragments on which we find the painted images of rayed sun disks and moon crescents. One example, dated to 1000 B.C., shows flowering edible plants accompanying water scrolls and solar symbols. This text could be a reference to the association between bright sunshine and rain and the germination and growth of the cereal crops.

The first recorded astronomical inscription in material made of bone has been dated to the 16th to 9th century B.C. Shang dynasty of Henan province. There are about 100,000 pieces of animal bone and tortoise shell containing an early form of Chinese characters, unearthed from what appears to have been a vast palace archive. Later historical descriptions of just how these materials were used, together with the partial decipherment of the calendars, implied that divination was the underlying motive for these astronomically related texts—hence the name oracle bones.

As at Delphi, anyone who wished to consult the oracle, usually a member of the royal family, would come with a question. In this case, the answer was received, not by listening after whispering the query into some godly conduit, but instead by looking at a visual pattern. The shaman would place a glowing piece of metal into one of a series of pre-drilled holes in a polished piece of tortoise shell. Then he would interpret the pattern of radiating cracks that resulted from the sudden expansion of the material. Full documentation of the prognostication would then be recorded in Chinese characters directly on the piece. For example, a mundane inscription might read: "The divination of date X (in lunar notation) was performed by Diviner Y. It will rain on date Z. And on date Z it did indeed rain." Note that unlike Babylonian prognostication a

This inscribed tortoise shell is just one item from a huge collection, unearthed from what seems to have been a palace archive, that has been dated to the 16th to 9th century B.C. The Chinese bone and shell engravings that carried astronomically related texts were used as a form of divination, and hence have become known as oracle bones.

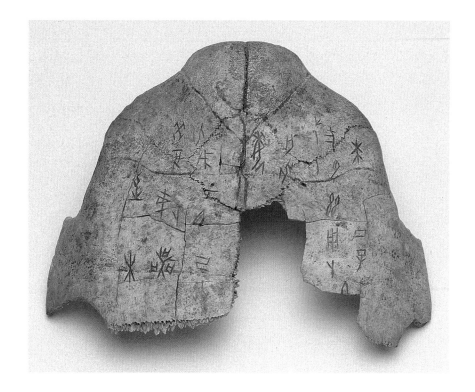

statement of confirmation is included with the omen, a credit to the attention to detail of Chinese record-keepers.

The vast wealth of astronomical detail recorded on the bone texts has enabled modern historians of astronomy, who can backtrack regularly recurring celestial events with the computer, to match sky phenomena with inscribed dates. Thus they can not only offer a fairly precise chronology of events, but also gain insight into how the ancient Chinese employed the information they collected. A cache of 5000 pieces of oracle bones excavated in Anyang, Henan province, in 1972 yielded a series of divinations of a single sky event widely referred to as "Ri-you-zhi." Chinese astronomy historian Zhang Peiyu found that all six dates recorded in the inscriptions matched perfectly with a series of solar eclipses visible from the Henan area in the 12th century B.C., half a millennium earlier than any records of such events obtained from Babylonia and Egypt. Other astronomical subject matter depicted on the bones includes novae, comets, stars that belong to several lunar mansions, and important planetary conjunctions. Some examples:

> On day *kuei-yu* it was inquired: The sun was eclipsed in the evening; is it bad?

> On day *keng-ch'en* it was inquired: There has been another *chih* [eclipse] of the sun and a sacrifice of nine oxen was made. Should this be reported to King Fu-ting?

On day *kuei-chou* it was asked: Will there be a disaster sometime in the next 10 days? On the day *keng-shen*, the moon was eclipsed.

The divination on day *ping yin* was performed by Kou. A sacrifice was made to the Fire Star (Antares).

Thousands of statements like these, usually beginning with a question, demonstrate the value of celestial events in foretelling the fate of king and empire alike. They give people a reason for watching the heavens closely and offer an alternative to the modern astronomers' quest for precise knowledge of the universe for its own sake. Shang astronomers were preoccupied with their heavenly dialogue, treating the starry denizens as though the determination of the future was quite beyond their own control. Like the Maya and Aztecs, these ancient people seemed ever mindful of the reciprocal balance sheet of debt payments owed the gods for their intervention in our earthly behalf. It becomes obvious that no matter to which culture we turn, we always seem to find astrology lying at the substrate of the earliest historical records that pertain to the sky.

We can attribute the incredible mass of observational records that emanate from China to a combination of astrology and good government. For centuries Chinese society has been bureaucratically organized. Family histories like those of the Jin dynasty (A.D. 265-420) contain lengthy chapters on astronomy. These consist mostly of astronomical records (where and when an object appeared and disappeared, its color, brightness, direction of motion) and the implications of these data in family affairs. There then follow lengthy discussions about the origins of astronomy, instruments, and sky concepts in general. Today, we would scarcely think of mentioning how brightly Arcturus glittered when putting together our family album! No merchant class like that of medieval Europe ever developed in China. Instead, the landed gentry of the early feudal system charted the direction of star and state. Unlike Plato and Aristotle, both of whom taught in a democratic city-state, Chinese philosophers were intellectuals of the court and the virtues they instilled bonded agrarian peasant class to ruling warlord and prince. There was little interstate scholarly communication, for science in China was not as outward-directed, basically, as it has been in the West.

What lesson did a Chinese peasant learn when looking high up toward the northern sky? He or she saw not simply a pair of wheeling bears flanked by a dragon, but a celestial empire, for Confucius compared the emperor's rule with Polaris. Just as the emperor was the axis of the earthly state, his celestial pivot was the polar constellation. The agrarian economy revolved around the earthly emperor the way all the stars turn about the immovable pole. According to one legend, the emperor was divine—born from the radiation of the Pole Star falling upon his mother.

Which stars did they recognize and why? China fixed itself on the pole to express its ideas about rulership. Four of our seven Little Dipper stars, plus two

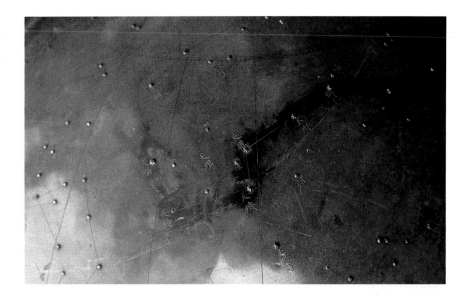

A bronze celestial globe in the grounds of the Purple Mountain Observatory near Nanjing was made in 1473 and is a replica of a 13th-century celestial sphere constructed by Chinese astronomer Guo Shoujing. The clearly visible Big Dipper familiar to Western eyes is known to the Chinese as "Pei Tou," or Northern Bushel—a scoop to measure grain.

others, constituted the Kou Chen, or "Angular Arranger," of the Jin dynasty. This seems an appropriate term for the celestial determinist. These stars made up the great Purple Palace, and each of their celestial functionaries had its terrestrial social counterpart (the emperor's palace). One member of the group was the Crown Prince, who governed the moon, while another, the Great Emperor, ruled the sun. A third, Son of the Imperial Concubine, governed the Five Planets, while a fourth was the Empress, and the fifth was the heavenly palace itself. So it went. When the emperor's star lost its brightness, his earthly counterpart would sacrifice his authority, while the Crown Prince would become anxious when his star shone dimly, especially if it lay to the right of the emperor.

Rulers have good reason for adopting stars of the north. Take Cassiopeia and Cepheus, the Queen and King of old Western star lore. They were also placed near the pole so that, like the power of the rulers, they could be visible eternally in the sky, never hidden by the horizon. The stars in the higher, temperate latitudes are raised quite high in the sky, and, as they turn about the pole, the fixity of the polar axis becomes a suitable metaphor among civilizations that developed in these locations. In the tropics, the appearance of the sky is entirely different.

The four surrounding stars of the Purple Palace proper are Pei Chi, the "Four Supporters." On the Chinese star map they appear well situated to perform their task, which is to issue orders to the rest of the state. The Golden Canopy, made up of seven stars, most of them corresponding to the wraparound, pole-centered stars of our northern constellation Draco, covered the palatial inhabitants and emissaries. Beyond them lay the conspicuous stars of the Northern, or Big, Dipper (Pei Tou). More concerned with realizing celestial principles here in the earthly realm, these "Seven Regulators" are aptly situated where they possess the maneuverability to come down and closely inspect the four quarters of the earthly empire. According to one version, the Big Dipper is the carriage of the theocrat who wheels around the central palace paying us periodic visits to check things out. Its stars establish the four seasons and distribute the five elements. They are the source of the yin and yang, the celebrated two-fold way of knowing that resolves the tension between opposing polarities (the male and female, light and dark, active and passive) that wax and wane through cosmic time to make up the potentiality of the human condition.

Second in importance to the polar region was the Chinese system (not strictly a zodiac, in that it was not confined to the ecliptic), which consisted variously of 28 or 36 houses arranged around the celestial equator, doubtless because the equator lies at right angles to the all-important fixed pole. This stands in contrast to the Western tradition, which uses the ecliptic. Traceable prior to the beginning of the first millennium B.C. when the zodiac may have been imported from India, the early choice of 28 is obvious, especially if, in a lunar-based calendar, you want to distinguish the position of the moon among the stars on a nightly basis strictly for the purpose of timekeeping. By following the nightly crossings of the meridian by each of the constellation houses, Chinese astronomers were able to track the course of the months within the 365.25-day year.

Of what do these constellations consist? Documents from fifth-century B.C. tombs give names of zoo- and anthropomorphic body parts like beak, mane, stomach, wing, heart, and gullet, while others are named after domestically related items such as harvester, house, well, ox, and winnower. Still others, like ghost and tristar (the three stars that make up the belt of Orion), seem more abstract. Whatever we may think, these names turn out to be neither obscure nor irrelevant to prognostications about the real world. The mane (or yak tail), for example, relates to events pertaining to warriors; the net to hunters; the lasso to prisoners, and the stomach to matters of the warehouse and granary. The (turtle) beak governs the harvesting of wild plants, while the ghost is capable of detecting cabals and plots against the emperor.

For every affair of state the starry winds of good and bad fortune would blow across the sky, and it was up to the astrologer to make the correct prediction. For example, one omen has it that when the stars of the well, which relate to all things quaffable, are faint, then wines better diffuse their aromas; but when its brightest star is especially red, then dark red drinks (and food) may be poisoned. Another example concerns the entry of a bright planet into the constellation called Mao by the Chinese (our Pleiades), the great sky sign of Tibetan warriors. An astrologer was thus said to have predicted the death of one of the most destructive alien invaders of the house of Tang. Two millennia after the tomb discoveries mentioned earlier, these same constellations appear virtually unchanged on Chinese projection maps that predate the Mercator projection technique of the Renaissance by five centuries.

Given the close parallel between the events surrounding the palace economy and the divinely ordained celestial arrangement, we will be no less surprised to find among the Chinese, as well as the Maya and the Egyptians, that royal architecture was erected so that it lay in perfect harmony with the land- and skyscape.

To assure that the situation of the royal capital would fit the local contours of cosmic energy, the king would summon his geomancer to perform the art of *feng-shui*. This expert used a variety of means—the nature of the local magnetic field, the consultation of oracle bones, the paths of streams in and out of the immediate environment and the categories of land forms—to help him precisely select and arrange a site. His secretaries would follow him around, taking meticulous notes as he performed his task. Sometimes workers would need to remove vast quantities of boulders or plant forests of trees to regulate the disposition of yin and yang energies that would pass in and out of the urban area. Paul Wheatley, a geographer and historian of religion, cites one such urban foundation ritual associated with the city of Luoyang of the Zhou dynasty of north China at the close of the second millennium B.C.

> In the third month, on the second [or third] day, Tiog-Kung (Chou-Kung: the Duke of Chou) began [to lay] the foundations and establish a new [and] important city at Glak (Lo) in the eastern state. The people of the four quarters concurred strongly and assembled [for the corvée] ... In the second month, the third quarter, on the sixth day *i-wei* in the morning the King walked from [the capital of] Tiog (Chou) and so reached P'iong (Feng). The Great Protector preceded Diog-Kung to inspect the site. When it came to the third month, the day *ping-wu* was the third day of the month. On the third day, *mou-shen*, the Great Protector arrived at Glak in the morning and took the [tortoise] oracle [as bearing] on the site. When he had obtained the oracle, he planned and laid out [the city]. On the third day, *keng-hsü*, the Great Protector and all the people of Ian (Yin) began work on the [public] emplacements in the loop of the Glak [river] ...

Louyang was not built in a day. As we can see from the attention to detail about place and time in this statement, acquiring proper urban form depended on getting the cardinal axes laid out precisely. This was one important task of the feng-shui expert, for the city needed to be accurately partitioned into quarters. It is written that when the constellation of Pegasus lay in the zenith, the diviner began to build the palace. And when he had calculated its orientation by the sun, he began to build the mansion. If we assume the time were sunrise,

The astronomical observatory in Beijing is depicted here in an 18th-century French engraving. The instruments on the roof show Jesuit influence as a result of the missionaries who visited China in the late 16th century.

OBSERVATOIRE DE PEKING
tiré du Pere LE COMTE.

the date would correspond to the June solstice. The shadow of a gnomon cast by the rising sun then would lie in the extreme southerly direction. By recording the shadow of the setting sun and bisecting the line between the two shadow tips, surveyors could arrive at the north-south axis. The method resembles the way the Egyptians precisely oriented their pyramids to true north. Here, however, the measurements could be supplemented by other data obtained with the magnetic compass.

Like Khufu's great pyramid, the walls of the Emperor's palace in Beijing today still line up perfectly with the cardinal directions that quarter the city. They offer a lasting space-bound reminder of the temporal schedule of duties of the emperor—to perform a specific task at the beginning of the first month of each of the seasons, these being determined by court astronomers who followed the course of the moon and sun, as well as the five planets, across the lunar mansions. For example, the emperor would go to the eastern quarter of his domain to start the new year every spring equinox. He would pray for a sound harvest; then, followed by his ministers, he would plow a ceremonial furrow in one of the fields. At the other seasonal pivots he would visit the remaining quarters so that everyone with a place in his empire's domain would also be accorded his own special time. This formal quadripartite seasonal calendar would have been familiar to any farmer, for it was based on what he could see in the sky. At the beginning of summer Antares lay due south at sunset, while at the outset of winter the tristar of Orion's belt took its place. When the handle of the Dipper pointed straight down, the peasant knew it was the first day of spring. Now was time for the king to come forth and speak to the people about the new year's harvest.

This working sundial is located in the courtyard of the Hall of Supreme Harmony, in old Beijing. As with all Chinese sundials, it faces toward the north celestial pole. The central rod, pointing in the same direction, casts a shadow that indicates the time.

In its high-level form, the keeping of observations and the preparation of the calendar all resided in the state observatory. Rather like the government research center at Los Alamos, New Mexico, or the headquarters of the Central Intelligence Agency, this top-secret institution lay hidden deep within the bowels of the Purple Palace. The importance of astronomical observing in the world of politics made such secrecy, even in the closest quarters, absolutely necessary. In the history of the Tang dynasty (A.D. 618-907), Joseph Needham, professor of Chinese history at Cambridge University, documents a directive issued by the king that reads: "If we hear of any intercourse between the astronomical officials, or their subordinates, and officials of other government departments, or miscellaneous common people, it will be regarded as a violation of security regulations which should be strictly adhered to. From now onwards, therefore, the astronomical officials are on no account to mix with civil servants and common people in general. Let the Censorate look to it." Another example of just how seriously the Chinese regarded the secret nature of astronomical knowledge is illustrated by a probably apocryphal, but nevertheless widely circulated, Chinese tale written as an epitaph on the grave of a pair of very ancient astronomers. They had paid a rather high price for bungling an eclipse prediction 30 centuries earlier.

> Here lie the bodies of Ho and Hi
> Whose fate tho' sad is visible,
> Being hanged for they could not spy
> The eclipse which was invisible.

Stressful and lonely, the ancient Chinese astronomer was hardly the free man of science we portray in his Western counterpart! Then, astronomers were nurtured by the government to perform a single, overriding task: to give the right time. But what *was* time for the Chinese? As far as we know, it emerges as a historically minded accumulation of seasons upon epochs upon eras. It begins with the mythical creation of life out of the breath of the divine architect craftsman who first chiseled the world out of the sky. It continues with the foundation of the original capital city and dynasty, and traces events down to the present-day festivities which celebrate all of these past happenings. Short time cycles were reckoned by the motion of the sun and moon, big ones by the five pacers (or planets) through the constellations, the stewards who escorted the brighter luminaries on their way.

Beyond sun and moon lay the planets and their cycles. Each member of the planetary pentad was identified with one of the five terrestrial elements: Mercury with water; Venus with metal (because it glowed "Grand White"); Mars, called the "Sparkling Deluder," with fire (because of its redness); Jupiter with wood; and Saturn with earth. But when it came to timekeeping, if Venus was the planet-watcher's fixation for ancient Maya astronomers of the New World, Jupiter was the clear choice in China. No one knows why, but I suspect it may have been the key to the lock on the commensurable cycles that caught their eye. Astronomer Edwin Krupp suggests that it may have been because Jupiter mimics the sun. It passes among the stars in 12 years the way the sun does in 12 months and it adheres closest to the ecliptic of any "wanderer" besides the sun. Also, Jupiter undergoes grand conjunctions in the same constellation with Saturn every 60 years. This is an exact multiple (5) of the time it takes for the 12 earthly branches, or *ti-chih*, of the equatorial zodiac to make a full circle. Place mats in Chinese restaurants still call to mind the popular names of the Chinese New Year: Rat, Ox, Tiger, Hare, and so on, which are designated by the magic 12. What may sound like number juggling to us has deep physical and moral meaning for those who fervently seek closed loops of time concretely marked out by divine celestial bodies. To an astrologer it would be quite natural to relate the 12-year cycle of the "Year Star" to the 12 earthly branches, especially if he wished to acquire a prediction about the political turf in whose celestial station a planet resided.

Omens as far back as the Shang oracle-bone texts (circa early second millennium B.C.) indicate that Chinese astronomers were onto the special cycle-bearing qualities of the planet Jupiter fairly early:

> The divination on day *hsin mao* was performed by Chi. 'The king is to make a sacrifice to Jupiter. Will it not rain?'

> The divination day *chi wei* was performed by Hsing. 'The king is to make a sacrifice to Jupiter. Will the offering of two oxen be sufficient to stop the disaster?'

Hundreds of years later the Jin family history gave an exhaustingly long list of the exact positions of Jupiter in various houses together with the sub-prefectures affected by each particular association.

The focus of Chinese planetary observations was on change and every change had a name and meaning. Each bend, kink, and turn, every slowdown or rapid motion the planets made along the bumpy celestial skyway was duly noted, for then their essences descended upon us. Particularly fascinating were the close conjunctions, especially the ones that took place during the retrograde segments of the orbits. To give a few examples, it is said that when the five pacers gather in the east, the signs affect the kingdom in the center and when they assemble in the west, the omens will concern foreign countries that threaten the center. When two planets lie in the same constellation they are said to be in combat. The closer together, the greater the magnitude of the calamity. If Jupiter and Venus enter into combat, the confusion is likely to be of a civic nature. Venus moving to the south and Jupiter to the north in close passage had its own particular designation, while the opposite passage had another name. One meant a good harvest, the other bad. Mars-Venus conjunctions had still other, quite separate, designations. And when three planets arrest their motion in the same place—then war will break out with a heavy death toll.

Given this listing of astro-social rules straight out of an old family history, we might be led to think that the business of Asian planetary astrology was totally objective and that every celestial gathering was evil. But we must not overlook the interpretive element, whether we believe it to be correct or incorrect, that attaches to all forms of astrology, both ancient and modern. The astrologer listened for the possibility of heavenly reprimands. His work always consisted of asking—and seeking answers to—questions. The objective element may be the position of a star in the sky; but the subjective aspect comes with interpreting, say, a pattern of cracks on the tortoise shell. And the answer always is given in the form of an increased or decreased potential for danger. Celestial news is not always bad, as we are sometimes led to believe. After all, if the messages from the stars were constantly deleterious, what need would any society have for the astrologer?

With the advent of the technological age, sky records in China became no less voluminous, only more precise. The history of Chinese instrumentation closely matches that of the West. Notched jade disks and cylindrical sighting tubes dating back to the fifth century B.C. probably functioned as a means of deriving and computing rudimentary forms of celestial cycles. We can document the gnomon in China as far back as 1500 B.C. By the time of the Han dynasty (206 B.C.-A.D. 220), an eight-foot-tall (2.4-meter-tall) version of a calibrated bamboo shadow-casting device was employed to mark noontime shadows in 1/10-foot (3-centimeter) gradations on a horizontal scale 10 feet (3 meters) long. As a result of measurements taken with this instrument,

The size and height of the buckets of this Chinese water clock, displayed in Beijing's Museum of Chinese History, are calculated so as to achieve an even water flow. A rod in the bottom bucket rises to indicate the interval of time passed.

85

This handsome copper-bronze armillary sphere, shown here standing in the court-yard of the Beijing Observatory, is now installed at the Purple Mountain Observatory in Nanjing.

astronomers needed to call a conference to decide on a calendar reform plan, so accurately had they fixed the occurrence of the summer solstice in the seasonal round. From then on the astronomers published an annual calendar. By adding rings for the ecliptic and celestial equator to this simple device, the Chinese erected their own version of an armillary sphere. By this time, Western influence had clearly spread throughout China in the wake of overland commerce from the eastern stretches of the Roman empire. Afflicted with the same megalomania and fueled by the same fervent desire to detect small variations in nature as their neighbors far to the west, the astronomer Guo Shoujing built an observatory tower at Gao cheng by the 13th century. Like ancient Athens' Tower of the Winds, this all-purpose observatory contained a number of astral measuring devices. The tower itself was the gnomon. A horizontal rod in an aperture at roof level, about 40 feet (12 meters) above the ground, cast a shadow on the wall below. A chamber at the top of the observatory was designed especially for watching the stars, while the inner rooms of the tower housed a water-driven clock and an armillary sphere.

While Europe experienced the Middle Ages with an inward-directed scientific focus of its own, Chinese globe and sphere builders came from a tradition steeped in progress. They were busy developing measuring instruments into cosmic models, many of which worked under their own power and consisted of complicated moving parts. This trend raises two important questions. First, what was the ancient Chinese concept of the universe at large? And second, how is it that Chinese thought never managed to create, out of

86

The observatory at Gao cheng in northcentral China dates from the 13th century, and was built by the leading astronomer Guo Shoujing. A horizontal rod in the upper aperture casts a midday shadow on the long, low wall which extends northwards from the base of the building. The wall therefore is a sky-measuring scale, and enabled Guo Shoujing to establish an accurate measurement of the length of a year.

new technology, an experimental science to test such models, the way the exact sciences evolved during the European Renaissance? The second question is especially intriguing since we know that in many aspects the Chinese possessed a superior technology to that of the West. For example, they had already invented gunpowder by the third century A.D., developed the clock escapement by the eighth century A.D, and invented the magnetic compass by the 10th century.

But first questions first. Just as we cannot understand the scientific Renaissance without giving credit to Islam, especially for giving us the language of mathematics in which to frame our quantitative speech, so too one cannot properly assess how Chinese astronomy materialized without giving due credit to India, itself a borrower and processor, for transmitting astronomical ideas and methods. David Pingree, a historian of Indian astronomy, has cited several different intrusions of Western astronomy into the Indian subcontinent, the earliest dating back to fifth-century-B.C. Mesopotamia and Persia, and second- and third-century A.D. Mesopotamia and Greece. But unlike the Moslems, the Indians, he argues, always strove to make the minimum possible alterations in the knowledge they acquired. They were more motivated to tinker with the mathematical details than reorganize the logical foundations and superstructure of the discipline. To repeat a rule in clearer language and to give a better example of its application—this was a far nobler act for an Indian astronomer than to speculate about the ultimate causes of things, for all celestial principles were god-given and never to be questioned.

This bronze incense burner in a courtyard of the Buddhist Lama Temple in Beijing portrays the universe as a cosmic mountain that penetrates the heavens and supports the celestial palace that occupies the zone of the circumpolar stars. This mountain corresponds to the polar axis, which seems to organize and mobilize the fundamental motion of the sky, and which imparted an equatorial character to Chinese astronomy.

We can understand the source of China's fixation with calculating time cycles by looking at the oldest Vedic texts from the India of 1000 B.C., which the Chinese inherited and which provide the strict and inalterable rules of timekeeping. Each *yuga,* or period, is subdivided into small cycles by year and half-year, month and half-month, down to individual nights. Then it is built up into inexperienceable super cycles longer than 4 million years, this interval being regarded as but a day in the life of the creator. As unreachable as these ideas may sound in the concrete world, the goal of Indian astronomy yet remained practical. The whole temporal set-up was geared to generate the times when people must perform their sacrifices to the gods. Like the ceque system of the Incas, here was a rigid scheme with a heavy dose of precise astronomy, all at the service of a complex and very ancient religious system. And all of it was taught and disseminated within the ruling family. Science was always a privileged intellectual activity, never to be conducted as a way of increasing public knowledge. After all, sacred subjects must remain confined within the ranks of the precious, qualified few who were honored to have been chosen as their skilled practitioners.

The earliest Chinese models of the universe came from India. As in the Vedic tradition, the earth is a circular disk, encircled in turn by alternating rings of land and water, all of it surrounded by the vast ocean. At the center is Mount Meru, at the base of which lies the Indian homeland. Mt. Meru's axis points to the Pole Star, conceived as the center of a series of concentric wheels, each bearing its own set of celestial bodies. Brahma uses the winds to turn the wheels. The sun advances successively along 180 different paths, each eclipsed repeatedly by the great mountain to produce the day-night cycle. Deviating wind currents pulled along by demons cause the planets to move in retrograde. The Chinese, who inherited these ideas, were not afraid to think in vast spatial terms. After all, as one modern astronomer has pointed out, they had no Greek concentric crystalline spheres to break out of. Consequently, they computed the sizes of orbits and the progression of planetary courses. One text for example gives the distance of the sun from the earth at the time of the winter solstice as 60,000 *li* and at its highest noontime position (summer solstice) as 80,000 li. One li is approximately 985 feet, or 300 meters, so this translates to a maximum distance of 15,000 and a minimum of 11,185 miles (24,000 and 18,000 kilometers), far short of the modern 93,208,277 miles (150,000,000 kilometers). Still, it is far larger than most ancient Greek astronomers, bound by their material spheres, ever would have contemplated. The sun's diameter was, incidentally, 1000 li (186 miles, or 300 kilometers).

From one family history we read: "The theory says that heaven ... hovers above the earth like an inverted basin. Both the center of the heavens and that of the earth are elevated while the water regions are low. Beneath the North Pole lies the center of both the heavens and the earth. The earth is highest, sloping down in all four directions..." Clearly, by this time the influence of the

West had affected the Indian-based model. While leaving the surrounding waters in place, cosmologists had mentally shaped the disk into a rotating sphere and surrounded it with a sky canopy. This late-model universe tries to relate the sky to the positions of the sun that can be determined from astronomical observations made with the gnomon. For example, it identifies the Tropics of Cancer and Capricorn, the zenith and horizon of the local place in a spherical universe. It even attempts to chart the course of the sun when it disappears from view. Obviously the model also cuts the bond between demon and wind.

Lastly, we return to the question of why China never produced a Tycho Brahe, Kepler, or Galileo capable of exploring ways of improving the models so that the facts of observation could be explained. I think the answer is that the cosmic process as we think of it was not in their purview. For example, the Chinese believed not in a single fashioning of the phenomenal world as we know it, but rather in a continuous, never-ending creation. Emptiness rather than materiality was their prime focus. Their creator was a maker of mutations rather than of things. Past, present, and future for the Chinese must be viewed in a combined whole that transcends whatever happens to be present in the flesh in the here and now. One cosmology text from A.D. 1000 will serve to illustrate:

> In the time of the Vast Prime (*hung yuan*) there was still neither Heaven nor Earth. Empty Space had not divided, nor had Clear and Turbid been separated within that obscure and empty desolation. The whole order of the Vast Prime extended into a myriad of kalpas (time cycles). After the separation of the Vast Prime there was the Composite Prime (*hun yuan*), and the whole order of the Composite Prime was for a myriad of kalpas and extended into a hundred fulfillments (even longer cycles), and the hundred fulfillments in turn were for eight-one myriads of years—and then there was Grand Antecedence. At the time of Grand Antecedence Lord Lao descended from empty space and made himself Master of Grand Antecedence.

It is difficult to understand the source of the power of conservatism in this statement about the cyclic nature of time that would sustain such record-keeping and leave sky concepts so unchanged through time as the record from India and China indicates. Keeping things in the family and confining science to secrecy because of the value of astronomical knowledge in political prediction certainly inhibited the sense of progress we value so highly in Western thought. But then can we justly believe that all civilizations would aspire to this attitude in science we call progress based on skeptical inquiry? And finally, we must remember that, except for the brief period of open sharing of the fruits of astronomy in ancient Greece, the notion of educating the public in the ways of progressive science is only a relatively recently acquired habit.

The pastoral, semi-nomadic Nuer, who reside in the upper Nile area, have two cyclic concepts of time. One is based on the changes in nature on social interaction and activity.

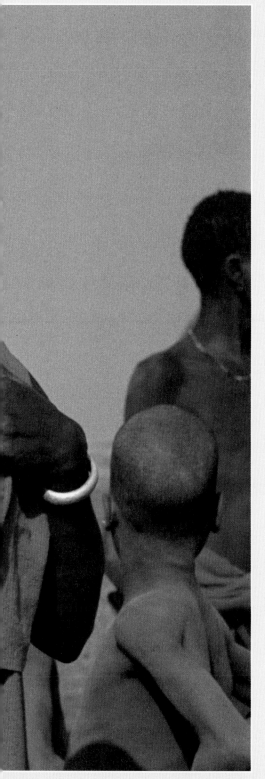

6

AFRICA'S SOCIALIZED ASTRONOMY

The Borana, Mursi, Ibo, Nuer, and Dogon—unfamiliar names of cultures that may be readily placed when classified as people who come from the countries of Ethiopia, Kenya, and Sudan. All devised calendars, all invented constellations and fabricated cosmologies that were based on practical observations and were especially well-grounded in principles of social cohesion. However, outside of Egypt, our knowledge of ancient Africa's first astronomers is scant. We must rely upon the relatively

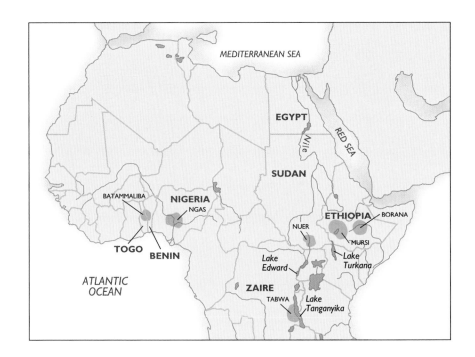

recent work of the first anthropologists to talk to people who had lived in relative isolation from the West. They discovered intriguing indigenous ideas and practices relating to the cosmos that probably were not so different from those expressed by the ancestors of these tribal cultures prior to European contact.

The Ngas of Nigeria still gather once a year for a fabulous moon festival. Priestly shamans decorate young boys with a white paste made of millet. One of them spits the decorative gruel on the boys' bodies, making a circular design upon their chests. These Sons of the Moon then go out to greet the new crescent by banging drums, shouting, and extending their long spears horizonward, as if to assault the pale luminary. And when the moon finally does come out, its apparition whispers heaven's message about the rains and planting.

The Ngas practice astronomy for the sake of the environmental and social activity that goes with it. But behind the Ngas' lunar ritual lies some serious astronomy that can be relevant for us, too, as we shall see. For how the moon tilts really *does* have a lot to do with predicting weather—a fact many of us have lost sight of because, unlike the Ngas, we no longer "shoot the moon," as they call it.

Oddly enough, although the history of the ancient Africans and their relationship with the sky is virtually a blank page, the first purported example we possess of recorded astronomical events may well have come from Africa itself. The Ishango bone was discovered in a small fishing village on the shores of Lake Edward, which straddles the boundary between Zaire and Uganda. The bone may have been part of a writing instrument, for when archaeologists studied it they found a small piece of quartz plugged into the opening at one

The Ishango bone (circa 6500 B.C.) was found near a lake located on the boundary between the present-day countries of Zaire and Uganda and has rows of notches. The bone may have been a lunar calendar.

end. Two rows of notches scratched on the bone's surface each tally 60, in one case splitting the count into 2 x 30, while a segment of a third row totals 30. Peabody Museum researcher and writer Alexander Marshack feels reasonably certain that the implement is part of a lunar calendar. But what would a family of Neolithic fishermen way up the Nile care about moon-watching? If you know anything about fishing the answer is obvious. Whether they frequent the tide-bearing oceans or shallow lakes, many fishermen today are well aware of the correlation between marine activity and the phases of the moon. They still gauge the time to go out on the water according to lunar tables.

Contemporary Borana people from East Africa developed a moon-star "timepiece," which gave them a way of keeping time through collective memory. They reckon the months of a lunar calendar by the times of appearance of the first crescent moon against a set of constellations that includes the conspicuous Pleiades, Orion's Belt, and bright Sirius, and the rather less conspicuous faint stars of our constellation of Triangulum. For half the year they commence each month when the first crescent moon appears in each separate star group, beginning with Triangulum—probably because it lies in a convenient location along the lunar course. For the second half of the year they look for the disappearing crescent in the pre-dawn sky in Triangulum and reckon the phases of the moon relative to that star group alone. They also mark the day by following the course of the moon among the stars irrespective of the lunar phases.

Just 155 miles (250 km) west of the Borana lives another skywatching tribe that had been fairly immune to outside influence—at least up until the disastrous drought of the 1980s. If you ask a Mursi tribesman what *bergu*, or month, it is, he probably will tell you that some people in his village recently told him it's the 5th, while others said that it was the 6th. Sometimes this dialogue about what time it is can become as heated as an umpire's close call at home plate or on the tennis court. This built-in disagreement about time had once been taken by anthropologists to imply that these semi-nomadic Ethiopians, who live by and depend upon the ecology of the Omo River north of Lake Turkana (Lake Rudolf), didn't care about marking time. What they failed to realize, however, as British anthropologist David Turton and astronomer Clive Ruggles later discovered, is that for these frequently mobile people, a calendar is quite unlike the thin numerical booklet of paper we hang on our walls. For them, keeping time is an interactive process, a dialogue among many people based on social rules that advocate an agreement to disagree, perhaps strengthened by the fact that the Mursi do not keep astronomical written records.

What motivates this need to argue and disagree about time? To answer we need to briefly analyze their system. The Mursi count 12 full moons in a season and every tribesman can recite the set of social and agricultural activities that go with the lunations or bergu. For example, in bergu 1 people select and move into riverside cultivation sites; in bergu 2 they clear the vegetation; in bergu 3 they plant sorghum, which they harvest in bergu 5. By bergu 7 they prepare the surrounding bushland for cultivation. The 8th lunation is the one in which the great transition from dry to wet seasons usually occurs, for then the lengthy rains begin. In bergu 11, after harvesting the latter crop, they start to gather honey. The 12th moon is one of celebrating year-end, a period of feasting and beer drinking. Now as we know, 12 moons add up to about 11 days short of a year. So the heated debate centers on the Mursi way of responding to the imperfect fit between lunations and seasons to which the Romans, we will remember, reacted by striking moon-time out of existence and substituting sun-time in its place. Following the last bergu, Mursi timekeepers created the period *gamwe*, an interval of activity that lay outside of the calendar. We are all familiar with ancient time-outs from the normal way of reckoning, like the extra five days at the end of the year in the Egyptian and Mayan calendars. Another example would be our familiar 12 days of Christmas—the leftover handful of days by which the seasonal year exceeds 12 lunations, which evolved from the old pagan calendar. Even our weekends can be thought of as time-outs from the normal course of things—hard work.

One difference is that in Ethiopia, gamwe is a full lunation in length, resulting in a 13-month year that exceeds the seasonal year by about 18 days, rather than a 12-month cycle that falls 11 days short. But how to recycle the year and make things fit? The trick is that gamwe is more subjectively determined than all the other bergu. It is tied to the flooding of the river and not to the more objectively classified phases of the moon. For the river will always crest in gamwe, and how high it crests will determine how extensive the flood will be. Though the Mursi reckon a gamwe in every bergu cycle, the floods occur neither at the same time nor to the same extent throughout Mursi land. What one person sees, another may view quite differently. Herein lies the answer to the question of motive. Unless people are willing to haggle over the calendar—even changing their opinion about what time it is—they will not be able to integrate all their valuable environmental observations and thus remain united as a wider community.

Astronomy is not separable from social questions among most African tribes. Nonetheless, Mursi flex-time is well informed by astronomical indications that fuel the fires of debate among the people about what time it is. For example, the names of the constellations, like those of the bergu, are tied directly to the agrarian activities on which they so vitally depend. The last appearance of the Pleiades in the west happens in bergu 9 and this star group is overhead at sunset in bergu 6. When Sirius and Canopus appear together in early evening the cattle must be returned to the homestead from grazing, for then it is bergu 5. In fact, the Mursi monitor the flood of the river by the stars. When the right-hand star of the Southern Cross (called *imai*) disappears in the evening solar glare, the river will have risen high enough to flatten out the imai grass which thrives along its banks. By the time Beta Centauri (*waar*) disappears, the Omo River, which the Mursi call waar, has risen to its maximum level of the low forest. The *sholbi* tree flowers along the Omo's banks when the star sholbi (Alpha Centauri) becomes obscured. These astral rules of thumb are a reminder of the many skilled observations of the Mursi's predecessors who farmed and raised cattle well outside the colonial sphere. They also give us a sense that for these people, time *is* activity and not some abstraction measured on a clock.

The farther south you go in Africa, the more your eyes are exposed to the splendor of the nebulous white light beyond the nearby stars that we call the Milky Way. In the nation we today call Zaire, on the shores of Lake Tanganyika, the Tabwa people use the entire Milky Way as an orientation-timing device. They call it "God's clock," and they give rough measures of it by hand, telling which way our Galaxy is oriented relative to the South Celestial pole, where no prominent Pole Star marks the point of celestial fixity as in the north. In the middle of the wet season in this area, the Milky Way lies east-west across the sky, while at dry season's mid-point it is arranged north-south. Stars within and astride the Milky Way have special names.

A young Mursi fisherman stands beside a river in Ethiopia. Mursi life is governed closely by the stars; the flooding of the two rivers between which they live is therefore relatively predictable.

Vega and Canopus, positioned almost exactly opposite one another, are close to the Milky Way, one just to the north and the other to the south. In dry season they are above the horizon and they also lie close to the celestial poles. This is why they each receive the name *Tulama*, or guardian of the pole. Orion's Belt, which is seen by anyone who ever has looked up on a winter's night, is assigned a similar representation in African as in Western star lore—he is a predator. The triad represents the tracks of the three-toed mythical aardvark-hunter who walks alongside and in the same direction as the Milky Way.

Astral knowledge of this type may have been preserved on pieces of tortoise shell to which were affixed bits of cowrie shell and beads. These devices, called *lukasa*, or the long hand, were in the possession of Mbudye, the Tabwa cult of ancestor worship. They may have functioned like the stick charts of the Marshallese of Oceania, as mnemonic schemes for recording astronomical knowledge. Perhaps they served in part as constellation maps; at least patterns depicted on some of them are said to resemble our constellation of Orion.

Why this Tabwa directional fixation? Could the emphasis on the north-south axis have some importance in navigation related to the basically north-south situated Lake Tanganyika, asks anthropologist Allen Roberts. We know that navigating the lake can be a difficult prospect. Or perhaps the "Wide Path" of the Milky Way was the legendary direction the Tabwa ancestors followed as they marched southward to a virgin land, having been driven out of

HOW TO PREDICT WEATHER BY THE MOON

We know that in a number of indigenous societies in Africa and elsewhere, people predict the weather by looking at the moon. But how do they do it? The graph below depicts the annual rainfall pattern in central Nigeria. As we can see, there is a wet season and a dry season. The appearance of the first crescent moon seen in the western sky after sunset also varies throughout the year in a regular way. Natives could note in particular the change in the way the points of the crescent tilt relative to the horizon. The diagrams of the moon below show the crescent on the first day of each month of the year as seen from the same place. Our graph shows what people could easily recognize by careful and repeated observation. When the moon begins to tilt right, the rains start and when the moon begins to tilt back to the left, the rains start to taper off.

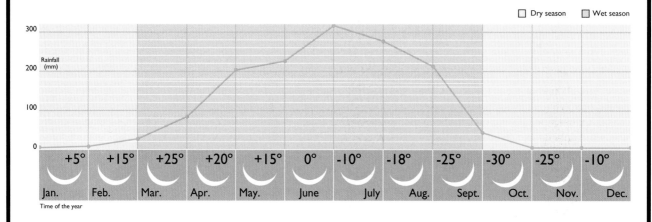

their original habitation by their ancient oppressors. The celestial opposition of this pair, along with the names of other stars, also has an astrological configuration. Sirius, for example, connotes predictions relating to betrothal. According to Roberts, it is called *Senda*, which means "to fetch" or "carry away," a possible reference to a woman's action when she moves to her new husband's village. Rigel's name, *Funzi*, or "to break everything," describes exactly what the violent spring rains do when this star sinks low in the west at sunset during this tenuous time of year.

The act of catching a glimpse of the first crescent moon was developed into a performance spectacle of theatrical proportions by the Ngas, who live on the Bauche plateau of Nigeria in West Africa. Anthropologist Deirdre LaPin has documented the ritual of "shooting the moon." Around harvest time the community of tribes from half a dozen local districts gather together to rid themselves of the impurities and misdeeds of the past season. Each tribe in rotation is given a month's assignment centered around the harvesting half of the year, which occurs from July to September on the high rocky plateaus. (The Ngas were fourth in the rotation when LaPin observed the rites.) In this assignment, a moon-timing expert is instructed to anticipate the night when the first thin crescent will be visible in the west by keeping records of the days of previous months by means of a series of knots tied on a piece of string. Such predictions would certainly be an acquired observational skill, for we will recall that sometimes a lunation can be 29 days and at other times 30—even 28 or 31 on rare occasions. On the day of the ceremony young boys are brought out to rescue the moon. Their faces and bodies are decorated with big white circles to symbolize the moon's light. A long line down each of their noses symbolizes the spears the elders will shortly fling in the direction of the horizon where the moon stands, as if to pin him down.

When the Ngas' turn comes in the monthly rotation, their chief and his elders, with the young sons of the moon and their calabash offerings in tow, go up to the observation site just below the high western ridge that overlooks the village. To begin his vigil, a chief squats behind a stone marker, reminiscent of the megalithic indicators used throughout the world by ancient astronomers. The moon expert says he knows where to position himself, always standing behind a rock marker and looking toward a particular valley between hills on the high ridge above him. All face the magenta-colored sky in twilight's growing dimness, and sing out to the moon: "Where are you? ... Lie on the pasture so that we can get you. We've already caught your brother...." Finally our astronomer spots the slivered, silvery disk. The men fire their spears and yell out a war chant. The boys shout down to the people below that the deed is done— they have shot the moon. Next day all the tribes gather to hold a funeral for the old moon. Like the bergu over the Omo River for the Mursi, the crescent moon of the Ngas is the vehicle of civic time in a socialized system of astronomy.

Why is the moon man's prey? What coordination among these villages exists to anticipate and locate the moon in its correct place? How far back does this elaborate ritual extend? We know from British documents that the oral tradition of the Ngas and their brethren can be traced at least to the 17th century. While astronomers and anthropologists have only begun to answer some of these questions, what we do know points in the direction of astronomy and moon-watching as a mechanism of social cohesion.

Many moon rituals have something to do with ancestor worship. Even among South African Hottentots, the cause of death always has been attributed to the waxing and waning of the moon. They tell the story of a rabbit whom the moon entrusted with an important message: When the moon comes back to life, so will the people. But the rabbit garbled the message, announcing instead that when the moon died, people would die too. When the rabbit returned to the moon and admitted his mistake, the moon punished him by hitting his lip with a piece of wood. Since that time the hare has had a cleft lip and it is forbidden to consume his flesh. If you look carefully you can still see him on the face of the lunar disk. Recall that the Maya of the New World also observe a rabbit in the moon.

The calabash offering in the Ngas' ritual also has a cosmological explanation. In fact, the calabash is a universal symbol throughout Africa. Even in Benin (formerly Dahomey) on the continent's west coast, the universe is said to be calabash-shaped, the horizon being the part where the upper and lower rims of a divided calabash come together. Stars move across the top half, while the earth is a smaller, flat-topped calabash that floats in a watery medium. If you dig into the earth you can witness some of the water from the underworld leaking through to the surface.

Like the Mursi calendar, the annual Ngas ritual reinforces the bonds of kinship among tribes who depend upon one another. Beyond this, moon shots have a practical value, for the local astronomer says that a knowledge of where the moon appears and the direction in which its crescent tilts offer a prediction about the crops. This is not superstition or fantasy. The box on page 96 demonstrates one way to predict weather—by looking at the tilting crescent moon.

Among the gods of many tropical African people the sun, who never had to be "called back" as in temperate latitudes, was always hot, fierce, and unrelenting. The moon was the moderating force, the half of the celestial twin deities to whom people could talk, the one capable of moderating the other forces of nature, the deity who alone had the power to regulate fertility. The moon lit up the dark where all the other spirits hid away.

In other African cultures the structure of the universe erected by the gods has a sunward directedness which is mirrored in the environment built by people here on earth—not just in the holy temple and the tombs, but even in the structure of their villages and domiciles. According to art historian Suzanne Preston Blier, the Batammaliba of Togo and Benin, inheritors of one of the great artistic dynasties of pre-contact Africa, are among the master builders of that continent, and their architecture has a decided cosmic theme.

Because the sun always has been present, these people reason that "no one, no power, ever could create this deity." And so to make their own houses as everlasting, they align its crossbeams carefully so as to point to the direction of the equinox sunrise and sunset, the butt end to the east and the narrower one to the west. Like the world calabash, the house is disk-shaped and features a raised floating terrace topped by a circular roof.

As in North America's Pawnee lodges, religion and cosmology intersect with astronomy in the Batammaliba domicile. "House-horns" located over the door refer to the number of days it took the creator to make the world, and conical mounds in front of the door commemorate the piles of stone used to build up the structure of the world. Newly built structures are consecrated by inverting a half calabash in front of the doorway, using it as a template to symbolically trace out the flat horizon-line-of-creation on the ground. A similar tracing is enacted around the entire Batammaliba village during a ritual in which the village "Earth priest" uses the feet of a sacrificed goat to circumscribe a line about the basically circular-shaped village, which consists of several dozen houses.

The Batammaliba, who live in Togo and Benin, align the crossbeams of their houses so that they point in the direction of the equinox sunrise and sunset. They believe the sun is human in form and lives in his house in the western sky. The doorway of the sun's house faces east; therefore all Batammaliba shrines must open to the west to face his home.

Unlike the Ngas then, the Batammaliba say the sun—*Kuiye*—is the one to talk to, for he created us and he protects us as well. However, we never see him—only the reflecting mirror positioned on his forehead. The real sun is human in form and he lies beyond the sky. His house is in the western sky, but the doorway faces east. This is why to commune with him all our shrines must open to the west to face his residence.

So far as African astronomy is concerned, however, here—more than anywhere else—we are denied the luxury of being able to document cross-cultural ties, such as we find in the more voluminous materials that have come down to us over the years with respect to the ancient Mayas or Greeks. These cultures possessed a form of expression and communication more familiar to us that we have labeled "writing." By way of contrast, astronomy in Africa emerges as a series of disconnected bits and pieces, the evidence shadowed in symbolic icons like the universal calabash and cosmic architecture, or in the very mundane eco-calendar where we see elements which we suspect have a deeply rooted native history. So much more work needs to be done on this continent.

An anthropologist once enviously said of the Nuer tribe of Central Africa, who possess a calendar very much like that of the Mursi, though based on seasonal stages in the life cycles of their cattle: "I do not think they experience the same feeling of fighting against time or having to coordinate activities with an abstract passage of time, because their points of reference are mainly the activities themselves ...." Above all, if we can attribute one distinctive quality to the people of sub-Saharan Africa who pay attention to the sky, it appears to be that astronomy and the things people need to do in daily life fit together harmoniously.

So concerned were the Maya with the influence that celestial events had over human affairs that they kept meticulous astronomical records Chichén Itzá, to chart the movements of the stars and planets. The windows at the top were used to sight the positions of the sun and Venus.

7

HERMETICALLY SEALED: ANCIENT MESOAMERICA

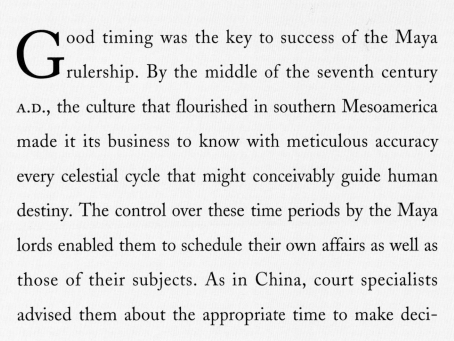

ood timing was the key to success of the Maya
G rulership. By the middle of the seventh century
A.D., the culture that flourished in southern Mesoamerica
made it its business to know with meticulous accuracy
every celestial cycle that might conceivably guide human
destiny. The control over these time periods by the Maya
lords enabled them to schedule their own affairs as well as
those of their subjects. As in China, court specialists
advised them about the appropriate time to make deci-

and built observatories, such as this one at

sions. When should we engage in armed conflict, set a date for a royal marriage, an accession to the throne, or conduct a ritual to pay the debt to the gods for their assistance in producing a good crop or a healthy newborn child?

Above all, submission to the will of the gods, the earth and sky deities who controlled the universe, lay at the foundation of all Mesoamerican skywatching, especially during the period from the beginning of the Christian era up to about the 12th century as measured by our time scale.

Our evidence relating to the New World's first astronomers comes from written books that contain hieroglyphs, carved monuments, and astronomically oriented buildings. Mesoamerican, as well as all other New World astronomical knowledge, was acquired without the aid of a sophisticated technology such as we have seen in Europe and the East. Evidence points to an enduring obsession with the celestial timing of human events that occurred in the everyday world. The Maya were fatalists at heart. They strove to find certain repeatable patterns recovered from the observations and recorded data of past sky events, which could then be used as a guide for predicting the future. For them, such patterns constituted realizable proof of the long-held Mayan belief that the future was present in the past—that the unfolding of events over time's near and distant horizon actually could be foretold by looking with introspection over one's historical shoulder.

We can be fairly certain that the habit of associating human events with moon phases goes back to a time that precedes sedentary civilizations. What is so unusual about the highly stratified, intensely specialized, agriculturally based societies of much of the Mesoamerican world is the degree to which models

By the mid-seventh century B.C., astronomers at various centers throughout Mesoamerica had achieved a high degree of accuracy in their predictions of celestial events, all without the aid of sophisticated technology.

Astronomers in different regions developed slightly varying methods for calculating lunar months. In Palenque, site of one of the great Mayan dynasties, they combined 43 months of 30 days and 38 months of 29 days. If this were computed in modern terms, a month would be 29.530564 days long.

for projecting moon-based time far into the future actually proliferated. And the Maya were masters of mathematically dialing up an enormous range of lunar rhythms that beat precisely in harmony with other natural periods.

Suppose that on the day when you first observe a thin crescent moon low in the west after sunset, you were to record that observation on a wall calendar with a heavy black marker. Now mark in red every day thereafter as you continue to watch the moon slowly wax to full, then wane again, and finally disappear as a thin crescent in the eastern pre-dawn sky. Suppose you mark the reappearance of the thin crescent in the west with a second black mark. If you were to repeat this process day after day and month after month, you would end up with a long series of red marks, each sequence representing intervals between successive, first-visible crescents. Between these sequences, there would be regularly interspersed a set of repeating black marks. This sequence, consisting of event-interval-event-interval, or black-red-black-red, reminds us of the basic structure of the Babylonian cuneiform calendars, as well as the controversial Mesolithic bone markings.

If you tallied up all the interval sets, you would find some to be 29 and others to be 30 days long, with a sprinkling of 28s and 31s throughout. This is because the synodic lunar cycle varies. At certain times and in particular places, when the apparent lunar orbit in the sky lies nearly perpendicular to the horizon, the moon's disappearance before new phase is hastened, while in other situations—for example when there is a low angle between orbit and horizon—first crescent can be delayed. Nonetheless, if you had the patience and developed a sharp eye for spotting the early crescent over years and decades, you would arrive at the conclusion that, no matter where in the world you happened to spot the moon, observable lunar months are easily remembered by a formula that consists of alternating 30- and 29-day months, about 55 percent of the former interposed among 45 percent of the latter. Such a formula would enable you to predict, with absolute certainty if society demanded it, what the phase of the moon would be on your just-born child's 20th birthday or on the centennial of your great-grandfather's ascent to the throne.

We know that astronomers of eighth-century-A.D. Palenque, in what is now southeastern Mexico, implemented a recipe that combined 43 months of 30 days with 38 months of 29 days; while those of Copán, farther to the south and east, mixed 79 of 30 with 70 of 29. Modern astronomy computes the length of the month of the phases as 29.530589 days, which is closer to 30 than to 29 as we might expect. If we computed the Palenque month in our terms, it would be 29.530564 days long, while the one at Copán would be 29.530201. Of course, such a time-averaged computation does not mean the Maya actually measured the lunar month to the third or fourth decimal place.

We have an idea about how the first astronomers actually arrived at these data. The process surely must have consisted of decades of tallying event-interval-event. The evidence that they did so comes from the lunar-based hieroglyphs that survive on the monumental inscriptions at the sites. One particularly dramatic testimony to perspicacious moon-watching survives in a Mayan written book called the Dresden Codex. It is worth discussing because it offers an excellent way to demonstrate the Maya penchant for finding natural cycles that are commensurate with one another. Rulers of the Maya dynasties employed such skills to connect the very roots of creation to astrological predictions about the future.

Though it was composed in northern Yucatán in the 12th century, the Dresden Codex is so named because it was discovered in the library of that German city in the 19th century, having passed through a variety of European hands since it likely was smuggled out of Mexico in the middle of the 16th century. There it lay among the many rolls and folds of parchment books confiscated by the holy men from the West who had been instructed to convert the pagan natives to Christianity. A Spanish priest—Diego de Landa, one of whose contemporaries may have spirited the codex away from Mexico in the first place—tells us something about those who used the codices, for what purpose, and the nature of their contents:

> They provided priests for the towns when they were needed, examining them in the sciences and ceremonies, and committed to them the duties of their office, and the good example to people and provided them with books and sent them forth.... Their books were written on a large sheet doubled in folds ... and they wrote on both sides in columns following the order of the folds.... The sciences they taught were the computation of the year's

months and days ... the fateful days and seasons, their methods of divination and their prophecies ... and how to read and write with the letters and characters ... and drawings which illustrate the meaning of the writings.

Unfortunately, the Spaniards burned almost all the records they found.

The Dresden lunar table was used to predict eclipses. It covers 405 lunations, arranged day by day for 11,958 days. Dividing the smaller number into the larger one, we discover that the "Dresden moon" would peg a single phase cycle at 29.5259 days. But a closer look at the table reveals that the moon's rhythm meshes with two other cycles, one related to eclipses and the other to a curious "sacred round" of 260 days unique to Mesoamerica. The entire document, like all Maya written mathematical presentations, is recorded in base 20, rather than base 10 system such as the one we use.

Before we discuss "why 260?" we need to digress briefly to inquire into why the Mayas used a base-20 system in the first place and how such a scheme was used to build up the lunar table. Our base 10 system surely must have evolved out of the habit of our pre-literate ancestors of counting on their fingers. But a full body count in the Mesoamerican tropics would have been composed of the fingers and toes, with sub-units consisting of hand and footfuls of whatever article was being counted. This is why we see dots (which represent ones) and bars (which represent fives) as symbols of quantity on the parchments instead

MAYAN NUMERATION SYSTEM

Dots represent ones, the bars are fives, while the closed fist, or stylized shell, stands for zero. At the top of the chart are the numbers from one to 19 and zero. Below, several examples of larger numbers are given. These numbers are tallied vertically in base 20. The position at the bottom is the unit position, above it is the 20 position, then the 400, with the 8000 position on top. The third column from the left, for instance, is figured as 1 x 400 + 5 x 20 + 18 = 518. When Maya counted days, however, the 400s became 360 and the 8000s became 7200s, etc.

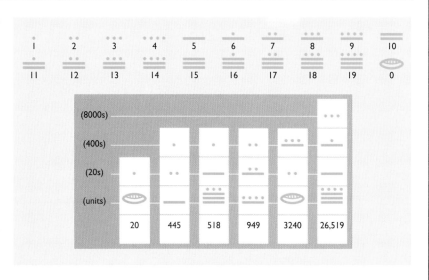

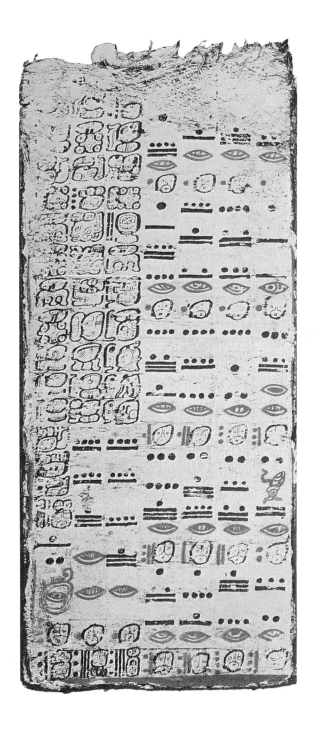
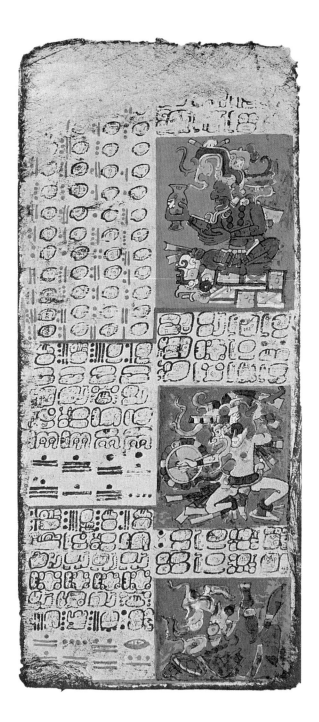

of the more familiar Arabic numerals. But like our very own Arabics, the Mayan symbols likely originated as hand gestures—ones being the tips of the fingers, fives the extended hands with the fingers together.

In base 20 every higher order in the system is cast in powers of 20 rather than 10. For example, in Mayan the number 8.2.0 would stand for zero ones plus two twenties plus eight twenty-times-twenties. Expressed in our decimally based system, this would be 3240. Note the zero in the lowest numerical order at the bottom in this particular notation (see box on previous page). Use of the zero gave the Maya an enormous computational advantage over their European counterparts, at least prior to its introduction in the Middle Ages via Islam. Anyone who doubts this might try adding or subtracting in Roman numerals. One essential difference in Mayan, just as in Babylonian notation, is that when time—as opposed to articles or things—is counted, the third place becomes 360, or 18 x 20, rather than 400, or 20 x 20; and the next higher unit becomes 20 x 360 or 7200 etc. This would have made the counting system more user-friendly when tallying units such as seasonal years. Thus, if 8.2.0 were intended to represent counted days rather than coconuts, it would translate, in our time currency, to 2920 days.

Given this brief mathematical primer on base 20, and before engaging the problem of where 260 comes into the picture, we next ask: Precisely what do the eight pages of the moon table of the Dresden Codex tell us? A chain of numbers across the bottom line of each page translates into a time packet of lunar synodic intervals. There are clumpings of six lunar synodic months (178 days) followed by one set of five (148 days). Each bunch of five moons is followed by a picture. A close look at all of the pictures together gives very strong clues about why the Maya would so parcel out a chain of 405 full moons over more than three decades. The answer is that Maya astronomers were attempting, apparently quite successfully, to predict eclipses. Some of these illustrations depict half-light, half-dark disks with lunar crescents opposing the *kin* glyph, symbol of the sun (kin means sun and day as well as time in the Mayan language). A serpent devouring the sun and a dead lunar goddess hanging by her hair from a segmented serpent who represents the sky also appear in the pictorial portion of the table.

Mayan astronomers were well aware that small cycles lead inevitably to bigger ones. Their codex seems to have been a mechanism for predicting not when the first crescent moons of the future could be sighted, but which full moons would be eclipsed and which new moons would eclipse the sun. It must have taken all of a century or more, which means several generations of perceptive astronomical observing, for specialists in skywatching to work through to a conclusion that their Chinese and Babylonian counterparts also had arrived at —that once a lunar or solar eclipse occurs, it is not possible to have another (of the same kind) until six, or more rarely five, months pass.

Of the long-range moon cycle in the Dresden Codex, this much is certain: First, it was used to gain control of astronomical time; and second, it was a time cycle derived from the observation of eclipsed and uneclipsed moons of the past, which could be used as data to generate a model for anticipating the occurrence of future eclipses—powerful knowledge in the hands of the rulership.

But why did the book's authors and copyists settle on 405 months as the basic choice in which to cast their picture-interval-picture/eclipse format? The answer is that ritual time, the other side of the coin of most ancient astronomical timekeeping, was as important to them as the time meted out by the stars. And this is where the sacred round of 260 days enters the picture.

One of the oldest time cycles in Mesoamerica, with roots dating in the archaeological record back to sixth-century-B.C. San José Mogóte in the Oaxaca Valley of highland Mexico, the count of the days (or the *tzolkin*) consists of 20 day names placed alongside 13 running numerals. The system worked rather like our familiar Sunday the first, Monday the second, Tuesday the third, a time scheme that pairs seven day names with (usually) 30 numbers. No one really knows how such a time loop got started. It is well known that 13 had long been held a sacred number (not a bad luck number) in Mesoamerica; for example, it represents the number of layers in heaven. And the significance of the number 20 has already been pointed out. What I think the Maya time-counters found so magical about the number 260, however, is that it fits together neatly with so many natural periods. For example, it is approximately equal to the average period Venus spends in the sky as evening or morning star (263 days); it is very close to the period of gestation of a human female (avg. 253 days); it approximately equals the average length of the agricultural season in many parts of Mexico. It even beats in the ratio of 2 to 3 with the long-term average interval between eclipses (the so-called eclipse year of 173.32 days). Natural coincidences like these give the 260-day round the advantage that once certain pairings of day names and numbers are assigned to eclipses, the same combinations will surface in future cycles; that is, eclipses will tend to fall only on or close to certain named days in the sacred calendar. All of this fits rather well with the Maya belief in lucky and unlucky days—the idea that, by their very designation, certain days embodied good or evil influences inherited from the past.

For all these reasons, the 260-day period emerged in the world of the Maya as the cycle par excellence to encapsulate the powers of all the gods—the gods of time, the sun, earth, moon, those of fertility, and rain. In a sense, 260 is the Maya divine temporal common denominator. What better way to make long-term projections than through commensurations of all the godly cycles with the ultimate cosmic time unit? Little wonder then that the 405 month length of the Dresden eclipse table is almost precisely a whole number of tzolkin (46 to be exact).

Other astronomical tables in the codices follow the same basic format: pictures mark events and numbered intervals record the duration in between the pictures. There is, for example, an ephemeris for Mars, another for Venus— even a table that charts the motion of the moon across the Mayan zodiac, which consists of a menagerie of very diverse animals not unlike the zodiac of Old World Egypt. Because of its extraordinary precision the Venus table is perhaps the most fascinating among all Mayan historical records. It runs through 65 synodic cycles of the brightest planet in the sky and encompasses a total of 146 tzolkin (about 104 years). A correction table that precedes the main text can be used to adjust the Venus table, rendering its predictive capacity accurate to within one day over five centuries. If you know how to use it, you can re-enter the chart precisely on the day name of Venus, which is 1 Ahau in the 260-day count of the Maya calendar. The table even manages to integrate the seasonal year in perfect step with the Venus cycle, thanks to the natural harmonic fit between the synodic period of the planet (584 days) and the length of the seasons (365 days). In effect, one measure of control of Venus time lay in the recognition that once the planet made a heliacal rise, it would do so again almost exactly on the same date of the solar year eight years later.

The events in the Dresden Venus table, which immediately precedes the lunar table, are the first and last appearances of Venus as evening and morning star. The intervals, written in red on the bottom line of each page of the table, add up to 584 days, but they are fractured into four sub-intervals in an unusual way. Astronomers assigned 250 days to the time Venus spends as evening star, while 236 are accorded to the morning star apparition. In between, 90 and 8 days mark the two disappearance periods. Now, the only one of these interval assignments close to reality is the last—the one immediately preceding each of the five Venus pictures. These show the Venus god in the act of hurling spears, which normally signify ill fortune in human affairs. Again, the choice of the too-long 90 days as well as the 236- and 250-day periods (the last two in reality ought to be equal at approximately 263) has an underlying logic. The Maya wanted to key the appearance of each of the Venusian stations to recognizable points in the lunar phase cycle. By so choosing these intervals, they thus guaranteed, for example, that Venus' first station (appearance) would correspond to the last date on which the morning star would still be visible when the moon was in the same phase it exhibited at the time Venus made its heliacal rise. This tendency to reckon changes in Venus's appearances with points in the lunar phase cycle may imply that the old Maya calendar, like most deeply rooted timekeeping systems, once had a short lunar rather than the much longer solar base line.

Like eclipse cycles and Venus appearances, Mars also possessed a numerical link with other known cycles. The Mars table, which precedes the Venus table in the Dresden Codex, exploits the unusual coincidence between a triple tzolkin (3 x 260 days) and the Mars synodic period (780 days). The planet is represented by a series of pictures of a long-nosed beast shown descending to varying depths from a sky band, the segmented body of a sky serpent that hangs over the scene. If the key event anticipated in the Venus table is its first morning appearance, then the comparable event in the Mars table appears to be the first perceptible backward motion and subsequent forward movement of the planet on its retrograde loop.

Given the quotations cited earlier from the Spanish chroniclers, there is little doubt about how these codex tables were actually employed. We can imagine the diviner to be skilled in both astronomy and astrology—for these were truly merged art forms in the Maya mentality. Thanks to the work of ethnologists like Barbara Tedlock we know something about how the modern version of the ancient Mayan diviner works. He pays a visit to his royal client, having been called from the court to make prognostications about civic affairs. Like modern diviners who still practice their craft in remote highland Guatemala, he sets his book and divining bag on a table. He draws amulets, beads, and crystals from the sack and arranges them in patterns through which he could count to set up some of the subjective rules for proceeding. Now the diviner determines when and where the celestial sign that brings the omen will appear. Is it a rising moon, a setting sun, or perhaps some phenomenon more difficult to pinpoint in time, such as a change of wind direction or heat lightning? As the king or queen posed questions about the opportune time to mount a raid on a neighboring city, the astrologer likely would have flipped through the pages of his Venus table, of which the Dresden version must have been but a local variant. Perhaps his Venus table was based on a standard set of tables, not unlike our nautical almanacs from which localized versions can be derived—like the old *Farmer's Almanack* that provides pertinent information for a New England farmer or sailor. But the Mayan books, though filled with marvelously accurate astronomical calculations, tell us little about the bureaucratic decision-making that attended the observation of sky phenomena. To understand what the people did when they actually witnessed these events—how they used their astronomical knowledge—we must turn to another kind of text.

That Venus was the patron planet of warfare for the Maya has been clearly indicated from the study of inscriptions carved on monuments, as well as in a rare mural painting from the ruins of Bonampak. About half the dates said to be connected with territorial battles depicted in the carved hieroglyphic writing on standing stelae fit with the very key positions in the Venus cycle we discussed above.

Furthermore, war seems to have taken place at Venus stations during certain months of the lunar count. The Maya observed a war-avoidance period during lunar eclipse intervals, for example. We have already noted in the Dresden pages that Venus's rhythm was written out in a kind of moon beat, which reflected the calendrical remnant of a lunar-based timekeeping system not unlike the one used in Islam. But why, of all planets, should Venus whet a warrior's appetite? In Old World mythology, Mars was the god of war, perhaps because his fiery red color reminded believers of the conflagration of one's material domain that often followed defeat in battle; perhaps, too, because Mars, like an ambitious warrior, covered all lands—the whole range of celestial

A set of murals in the Great Palace at Bonampak commemorates the victories of the eighth-century king Chaan-Muan, seen here with a group of captives in supplication at his feet. Four constellations overhead give clues to the celestial context in which this war-related scene takes place. Venus, the Maya planet of war, appears in various houses of the zodiac. Venus glyphs, for example, are depicted on the back of the tortoise at upper right and on the peccaries at left. Some scholars interpret these glyphs simply as stars.

turf in rapid order, from the side of the firmament occupied by the sun to the remote dark reaches of the midnight sky. While we can chart the planet as closely as they, we cannot conjure up the mental imagery that connected the appearance of the bright white light of Venus with the act of battle.

Deep in the Maya rain forest, Bonampak straddles the border between Guatemala and the Mexican state of Chiapas. Brightly colored murals, one depicting a dramatic battle scene, decorate the walls of the Great Palace there. The scene that commemorates the battle is about 15 feet long and 9 feet high, and contains more than 100 figures clad in full regalia. At the center, larger than life size, stands the early eighth-century king of Bonampak, Chaan-Muan, clad in a jaguar skin. Below him, on the steps of his palace, his worn-out captives appear, looking as if they had just been dragged in from the battle-field. Weary and humiliated, they shed drops of blood from their hands at the king's feet as they prepare to endure further punishment. Will they be beheaded or spared and returned to their home city? Perhaps artists from the losing side will be required, as a form of humiliating punishment, to come to Bonampak to paint the very scene we witness in the mural. Above this scene of supplication lie four depictions of constellations, among them a tortoise, which

Many Mayan rulers seem to have adopted patron planets and at Copán this can be witnessed in the impressive Temple of Venus, which was built during the reign of a ruler we call 18 Rabbit by the glyphic signs that make up his name. The facade was adorned with a double-headed sky serpent with Venus glyphs. It also contained a special slotted window through which the ruler's astronomers could time the evening appearance of the planet in the west.

may be identified with our belt of Orion, and a peccary, which may have corresponded to another star group not very far away in the sky. Certain of these signs are similar to the zodiacal band that appears on a page of one of the Mayan codices and on the friezes of other Mayan buildings. On the back of the tortoise and arrayed on and about the other constellations are Venus hieroglyphs, the very same ones we see written in the Dresden Venus table.

Was the attack real or just propaganda? Did it actually happen, and was this military iconography tied to the place where Venus appeared in the sky at the time of the battle? If so, could not the enemy, presumably as wise in the ways of astronomy as Bonampak's skywatchers, have cracked the secret star-wars code? It could scarcely have come as a surprise, for all raids were conducted during specifically designated war seasons, timed so that there would be crops, necessary to feed an army of foot-soldiers, standing in the fields. Also, it had to come during a season when intensive agricultural labor was not required. At least this is the indication on the seasonal war-related dates carved on Bonampak's stelae. Practically all of these dates can be pinpointed during the calendar period from November to mid-February. Heliacal rising and setting are the timed events in the Venus cycle that fit the monumentally inscribed dates. But some of the dates also match Venus's greatest apparent distance, or elongation, from the sun, approximately when the planet stands highest in the sky at dawn or dusk—or, more precisely, the first perceptible falling motion from its celestial high point. Perhaps it was the turning point or pivot in the planetary cycle that signaled the battle event. Who would have believed that by studying Mayan astronomy we would learn that their concept of warfare differs radically from our own?

Mayan rulers seem to have adopted aspects of their patron planets with the same enthusiasm that modern sports fans follow their favorite teams. For example, one of Copán's kings, Yax Pac, or First-Sun-At-Horizon as his hieroglyphic name translates, seems to have favored morning over evening star Venus appearances to schedule his raids on nearby cities such as Quiriguá. His grandfather, 18 Rabbit, had adopted the same procedure two generations earlier. 18 Rabbit had even built himself a temple of Venus. Its facade was adorned by a double-headed sky serpent with a Venus glyph at one end and a sun sign at the other, perhaps mimicking in stone the sinuous movement, lasting several weeks, of the imaginary line that connects the sun to Venus in the sky. Moreover, the temple had a special slotted window in its west facade through which 18 Rabbit's astronomers carefully timed the evening appearances of the planet. Copán's Temple of Venus was, in essence, an astronomical almanac in stone. Modern Yucatec-speaking Maya have told visiting anthropologist John Sosa of an old myth about a two-headed serpent who lives underneath the world. Every night the serpent positions himself to swallow the descending sun, which his other head then disgorges on the eastern horizon at dawn.

Recently, the inscriptions of Yax Pac's father, Smoke Monkey, have been deciphered. According to epigraphers Linda Schele and Barbara Fash, all of Smoke Monkey's Venus-related inscriptions are a mirror image of his own father's, 18 Rabbit. Smoke Monkey seems to have hitched his affairs to Venus, but not as evening star. Instead, he timed the dedication of his palace to occur two Venus rounds after that of his predecessor, but he based it on a morning rather than an evening star event. He even closed off the west-facing window in his father's temple. But why would a son revert to the celestial habits of his grandfather? Perhaps even kings are no exception to the dictates of human behavior, and in every child there may exist an impulse to be different from his parents. In the serious world where royal Mayan history and astronomy coincide, young Smoke Monkey's measure of independence seems to have been expressed in selecting certain aspects of a celestial body, each representing particular powers of ruling deities.

While the Venus cult flourished at Copán, far to the west in the city of Palenque, the late seventh-century-A.D. ruler Chan-Bahlum—his name means Serpent Jaguar—expressed a special fondness for Jupiter. Perhaps because he had succeeded a very famous father, Pacal (Shield), who ruled for several decades at Palenque and constructed its most elaborate temples up to that time, young Chan-Bahlum needed to distinguish himself. Whatever the reason, the events in his life seem deliberately attached to movements of Jupiter, specifically to the two stationary points of its retrograde loop. For example, his heir-designation ceremony, accession to the throne, and his apotheosis after death, all fell within a few days of a second stationary position in the Jovian retrograde loop.

Jupiter seems to have been the planet favored at Palenque during the reign of Chan-Bahlum. He is seen here to the right of the Mayan tree of life on a stucco plaque from the Temple of the Cross receiving the symbols of royal power from his dead father, the great leader Pacal. A rare planetary conjunction of Mars, Jupiter, and Saturn was interpreted as sanctifying Chan-Bahlum's dynasty.

One of the most spectacular of all celestially based events recorded in Mayan monumental inscriptions has been attributed to Chan Bahlum. It happened on July 20, in A.D. 690, the date called 2 Cib 14 Mol in the 52-year calendar round. (The first half of this date is the position in the 260-day cycle while the second half, the 14th day of the month of Mol, gives the position, by day number and month name in the 365-day year, which consisted of 18 months of 20 days plus an added short month of 5 days. Identical combinations on the 260- and 365-day "time wheels" occur at intervals equal to the lowest common multiple of 260 and 365. This is equal to 18,980 days or 52 years of 365 days.) Thanks to modern astronomical tables, we know that on this night a

rare planetary conjunction took place. It involved Mars, Jupiter, and Saturn, which had been dancing about the night sky ever more closely together. The date is inscribed in several texts on the stuccoed plaques at Palenque's ruins. Each inscription seems to imply that the three major sky gods responsible for the most recent cyclic creation of the world, which is stated to have taken place at Palenque in their equivalent of 3114 B.C. (our time), were re-assembling in the sky over their special city. Their appearance would reaffirm the descent of Chan-Bahlum's dynasty through the bloodline of the gods themselves. Thanks to his astronomers, what better way for the new king to legitimize the continuation of his father's dynasty than by staging such a rare celestial spectacle in the plaza fronting the temples where the inscriptions are written? It all came at a most appropriate time, for Chan-Bahlum had big shoes to fill.

How shall we view the motive behind this public exhibition of such a vast quantity of astronomical information connected with major events in the life of the Maya city, and especially its rulership? Perhaps as a way of justifying their wealth, status, and power, Maya rulers seem to be telling the world, through their public carvings, that they were indeed capable of seeing into the future. Only then could they and their subjects prepare for the practical as well as ritual activities that directly manifested their dialogue with the gods of nature, a philosophy that most members of the Maya citizenry, upper as well as lower class, must have taken quite seriously.

Just how far back can we trace this habit of event-scheduling by the stars in Mesoamerica? There are archaic clues to the nature of Mesoamerican sky-watching in the pictograms and petroglyphs that appear on the rock outcrops surrounding nascent ceremonial centers in Northwest Mexico. As we have seen, tally marks denoting the months and years were made by hunter-gatherers who spent but a handful of moons in any one location, depending upon whether or not their memory told them that the deer were running or that the berries would beckon ripely on the bush. Like the relatively small, kin-related units who roamed prehistoric Britain's Salisbury Plain, once these itinerant people began to band together in larger groups, and especially once they became sedentary agriculturists, they honed their relatively simple, arithmetical tallying into more carefully crafted calendrical computations.

Some of the earliest evidence that the first Mesoamerican astronomers had devised a sunwatcher's calendar is based not upon a written record, but rather on a more subtle form of expression. Along the Pacific coast of Guatemala, between El Salvador and the Mexican state of Chiapas, archaeologists have unearthed the remains of hundreds of Pre-Classic mounds dating to the early first millennium B.C. The discerning eye can pick them up every few miles or so along the Pan-American Highway. Each site consists of up to a dozen or more grass-covered, artificial hills, some of which once served as temple platforms, others as bases for the houses of the wealthy. At the center of each mound cluster we find a complex consisting of four large mounds arranged in the shape of a four-leaf clover. If you stand on one of the mounds on the shortest day of the year and look across the open plaza to its counterpart located 180 degrees opposite, you will notice that the line of sight often marks the place where the sun rises. This habit of incorporating the start of the year at the winter solstice within the architectural plan of a ceremonial center signaled the beginning of a tradition of astronomically aligning Mesoamerican cities. Thus began the idea of incorporating divine celestial events within the very structure of the habitation environment—a trait that became more elaborate as the urban civilizations of Teotihuacán in central Mexico and Aztec Tenochtitlán (site of modern Mexico City) developed—centuries, even millennia later.

The Aztecs of Tenochtitlán, who visited nearby ruined Teotihuacán during the time of Europe's Middle Ages, said that it was the birthplace of the gods. Already abandoned for several centuries, Teotihuacán had been built before the time of Christ. But this city was not constructed helter-skelter, the way medieval European cities evolved. Archaeological evidence suggests that it was carefully pre-planned and that, like Mecca and Jerusalem, it was as much a holy place as a center of commerce. Along its Street of the Dead, in the stucco floor of a building located not too far from the great Pyramid of the Sun, we

find eroded yet visible evidence attesting to the precise course taken by its architects. More than 2000 years ago, they had surveyed and laid out the 50-square-mile (128-square-kilometer) ceremonial center whose periphery would come to house more than 100,000 people. The evidence consists of a design made up of pecked marks in the shape of a double circle centered on a cross. It matches almost exactly another design carved on a rock outcrop a mile and a half to the west of the sun pyramid. In the 1960s archaeologists discovered that a line between the pair of benchmarks is almost exactly parallel to the east-west street of the ancient capital. A third marker high on a mountain to the north that overlooks the city, and a fourth on the south marked out other significant geographic directions. It is likely that these benchmarks were put there to establish that the city was an integral part of the cosmos. Only those directions deemed valid by divine will should be reflected in the domain the gods had once established here on earth—the home of the rulers of Teotihuacán, who descended from them, and the Aztec polity which believed itself to have descended from the great city.

But the east-west axis of Teotihuacán yields an even more complex temporal marker than the mounds that lined up with the solstices and which had been erected by their Pacific Coast predecessors. If you stood over the marker on the Street of the Dead 2000 years ago and cast your eye over the petroglyph on the western horizon at the correct time of year, you would have seen the prominent little star cluster we today call the Pleiades, setting. Because the stars have shifted their rise-set points over time, we can no longer see this ancient event in precisely the same place. Nevertheless, as has been demonstrated in a modern planetarium, when the markers were erected, the alignment with the Pleiades at Teotihuacán was nearly perfect and it probably had a two-fold significance. First, our Seven Sisters (as the star group is known) passed directly overhead in the latitude of Teotihuacán, thus signaling the fifth cardinal direction. And second, when the Pleiades reappeared in the east at their heliacal rising, after having been lost in the light of the glaring sun for 40 days, they showed themselves again on the very day the sun also passed the zenith.

Here was a highly visible, convenient timing mechanism to signal the start of the new year. Tying the sun to the stars, this temporal device was quite different from commencing the solar calendar by marking the sun's most northerly or southerly passage. The Pleiades, being both prominent and in the right place at the right time, became the new celestial timer of choice to these highly innovative astronomers who advised the lords of Teotihuacán. But this new Teotihuacán calendar was more natural because of another natural coincidence. Its tally system easily accommodated the body count, for the city lay at the precise location where exactly twice the body count in days (i.e., 40 days) also marked the period between sunrise on the cosmic axis between the benchmarks and the day of the observed solar zenith passage.

This magical orientation, approximately 15½ degrees to the east of north and west of south, was copied all over Mesoamerica for several generations after the fall of Teotihuacán. Dozens of petroglyphs resembling the Teotihuacán pecked circles have been discovered by archaeologists at sites ranging from the far north of present-day Mexico to the remote, southerly Mayan ruins of Guatemala. In fact, iconographic evidence uncovered by astronomer John Carlson suggests that even the Maya habit of conducting star-wars and raids based on the favorable positions of Venus may have originated in the city of Teotihuacán.

Registering the appearance of sun, moon, planets, and stars in the doorways of specialized buildings, as we have seen, was rather widespread throughout ancient Mesoamerica. Even the 16th-century Spanish conquerors had heard about the practice when they entered the Aztec capital of Tenochtitlán, which lies just 20 miles southwest of Teotihuacán. One informant told Spanish friar Toribio de Benavente Motolinía that a festival took place when the sun stood in the middle of the great temple at the equinox; but because it was a little crooked the great Aztec chief, Moctezuma II, needed to pull the temple down and straighten it.

The temple in question—called the Templo Mayor by the Spanish invaders—is the most famous of all the ancient Aztec buildings in ancient Mexico City. It was thoroughly re-excavated in the early 1980s after underground electrical workers, engaged in installing one of the city's new subway lines, accidentally broke into what proved to be an offertory cache of jades, decorative shells, skulls, and flint knives that had been placed there 500 years earlier. Subsequently, archaeologist Eduardo Matos Moctezuma directed the excavation through seven nearly identical buildings built on top of one another and spanning two centuries. He exposed each facade, so that the alignments now could be measured. The resulting east of north by south of east orientation is just what it would have had to be to permit the rising equinox sun to fall into the notch between the twin temples that once surmounted the flat-topped, 150-foot-high pyramid. When the sun reached that point, Spanish chroniclers tell us, a royal observer situated in the plaza fronting the bottom of the stairs carefully watched it. Like a town crier, he signaled the time to begin a ritual of sacrifice that attended each month of the year. Bernardino de Sahagún, Spanish priest and author of a 12-volume work on the history of Aztec Mexico City tells what happened during one of these rituals:

In the stucco floor of a building near the Pyramid of the Sun in Teotihuacán is a peck-marked design, like the one shown above. It has the shape of a cross surrounded by a double circle and it matches others to the west, north, and south of the site. These were more likely used to establish that the city was in perfect arrangement with the cosmos.

The Templo Mayor, shown here in model form, is one of the most famous of Aztec buildings. When it was erected in the ancient city of Tenochtitlán it was positioned so that the rising equinox sun would fall in the notch between the twin temples that surmounted the flat-topped pyramid. One of these temples was dedicated to the rain god, the other to the god of sun and war. When the sun reached that point, a royal observer, positioned in the plaza opposite, would signal the time to begin a ritual of sacrifice.

... the debt-payment which they celebrated during all the days of the month ...took place, then the feast day was celebrated for the Tlalocs.

There was the paying of the debt [to the Tlalocs] everywhere on the mountain tops, and sacrificial banners were hung. There was the payment of the debt at Tepetzinco or there in the very middle of the lake at a place called Pantitlan.

There they would leave the rubber-spotted paper streamers and there they would set up poles called *cuenmantli*, which were very long. Only on them (still) went their greenness, their sprouts, their shoots.

And there they left children known as 'human paper streamers,' those who had two cowlicks of hair, whose day signs were favorable. They were sought everywhere; they were paid for. It was said: 'They are indeed most precious debt-payments. [The Tlalocs] gladly receive them; they want them. Thus they are well content; thus there is indeed contentment.' Thus with them the rains were sought, rain was asked.

And everywhere in the houses, in each home, and in each young men's house, in each *calpulco*, everywhere they set up long, thin poles, poles coming to a point, on each of which they placed paper streamers with liquid rubber, spattered with rubber, splashed with rubber.

And they left [the children] in many [different] places.

This rite of spring, initiated by a sunrise over the Templo Mayor, served to pay the debt to Tlaloc, the rain-bringer, the same fertility god whom the ancient Teotihuacános had worshipped a millennium before. This was the time of year to watch the clouds begin to form over the mountain home of the rain god. This hill (called Tlalocan, or house of Tlaloc) on the eastern range of mountains confronting the temple precinct, is still the place where the dark clouds that bring the summer rains start to assemble. So there is logic to the Aztec myth that all water comes from the inside of the great hill where the rain god lives. And it makes sense, too, to offer the precious tears of the youngest members of Aztec society as an inducement to the rain-bearer during the most youthful part of the year—its first month.

It also seems reasonable for the Aztecs to line up the base of the great temple not only to face the sun—manifested in the form of Huitzilopochtli, the god of sun and war, in the left of the twin temples on top—but also to point toward the home of Tlaloc, to whom the right of the twin temples is dedicated. Remains associated with the ancient ritual of paying their debt to the rain god still stand atop Mount Tlaloc. A 300-yard-long causeway, built of rough-cut stone, leads to a ceremonial precinct where the Náhuatl-speaking descendants of the Aztecs still come to worship a Christianized version of their ancient deity. A child is said to have been sacrificed to the god of rain there as recently as 1889.

The Pleiades had survived as a recognizable Aztec constellation likely descended all the way from Teotihuacán. They called the constellation the market place, perhaps because of the closely gathered nature of its component stars, and Sahagún tells us that in the 16th century their precise sighting determined not just when to start the year, but also when to begin a calendar round—the same 52-year cycle we talked about that made up such an integral part of the Mayan calendar.

According to Sahagún, the ceremony of the Binding of the Years took place every 52 years. It began when the Pleiades crossed the overhead position at midnight (about mid-November), a statement that suggests the Aztecs were marking the time of night. When the time approached, the priests ascended the Hill of the Star to watch the movement of the Pleiades with great anxiety:

> And when they saw that they had now passed the zenith, they knew that the movements of the heavens had not ceased and that the end of the world was not then, but that they would have another 52 years, assured that the world would not come to an end.

While Sahagún identifies the Pleiades in his informant's drawing, made only a generation after the Conquest, it is unfortunate that the prelate did not specifically tell about the other star groups that also appear in the drawing. One of them could be the same Scorpion we identified in the Maya zodiac, while the meaning of some of the other pictures, for example there are some with the sun and moon faces joined, seem fairly obvious. In the volume of his works dedicated to astrology, Sahagún also tells us that Venus was associated with a blood sacrifice; warriors spiked their arms and legs with thorns, then flicked their blood at the newly-arrived morning star, he says.

Each of the marks tallied on a piece of barn door in Chamula, Mexico, by a modern day-keeper stands for one day. Heavy marks appear at 20-day periods, the entire tally making one year of 18 months x 20 days plus one month of 5 days.

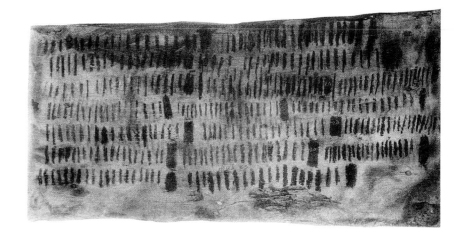

The more we examine the surviving evidence on Mesoamerica's first astronomers, the more we realize that it tells of a single, continuous culture, not the disparate collection of wandering, disconnected tribes so frequently portrayed in much of the older literature. The Teotihuacános seem to emerge as the New World's ancient Greeks. Perhaps we can think of the Maya world as the symbol of cultural Renaissance and the Aztec as embodying the last attempt in this part of the world to organize skywatching in the setting of a great imperial state. Soon, much of their world would be destroyed by European invaders seeking gold and precious stones, not to mention fresh souls for Christianity. Prior to the contact with Europe—a full generation after Columbus's first voyage—these cultures had thrived for centuries in total isolation from the rest of the world, hermetically sealed by two vast oceans.

Today, Mesoamerican culture is not dead, it is only transformed. Many native dialects, myths, and religious beliefs remain alive, though they have receded into the ever vanishing remote areas far from the byways of tourism. About 20 years ago, American anthropologist Gary Gossen happened upon an old woman using a piece of charcoal to scratch black marks on a fragment of a barn door. He was astonished to discover that the ancient form of the calendar we talked about in this chapter was still being kept—this despite the fact that the European calendar had been in full swing, even in such a remote place, for more than three centuries.

The calendar board found by Gossen in Chamula, a small town in the vicinity of San Cristóbal de las Casas in the Mexican state of Chiapas, is a far cry from the subtly intricate Mayan calendar we analyzed in the pages of the Dresden Codex, but enough remains on that old barn door to prove that deeply rooted beliefs can survive any ideological purge no matter how forceful the bearer of new and radical thoughts from far across the sea.

The Navajo custom of painting with sand was a means of depicting what these early Americans saw in the sky and also a way of communicating produced today during healing ceremonies as part of the ritual for attracting a particular healing power.

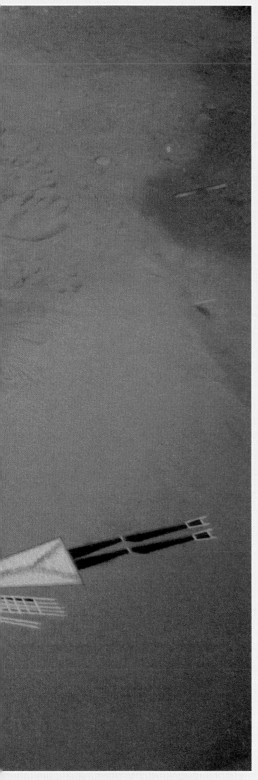

NORTH AMERICA: CHARTING CELESTIAL SYMMETRY

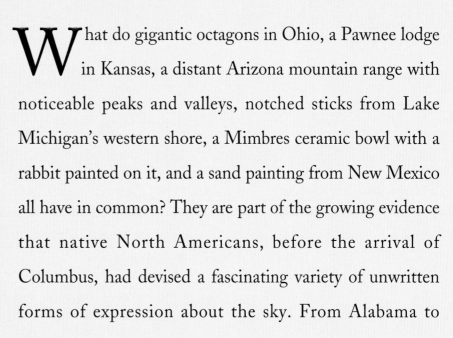

What do gigantic octagons in Ohio, a Pawnee lodge in Kansas, a distant Arizona mountain range with noticeable peaks and valleys, notched sticks from Lake Michigan's western shore, a Mimbres ceramic bowl with a rabbit painted on it, and a sand painting from New Mexico all have in common? They are part of the growing evidence that native North Americans, before the arrival of Columbus, had devised a fascinating variety of unwritten forms of expression about the sky. From Alabama to

Arizona to North Dakota, they recited star stories to back up what they saw. The Adena-Hopewell people of the eastern Great Plains built astronomically aligned abstract figures of perfect geometric proportions. And the largest mound by volume in the Americas, star-crossed in its configuration, lies not in Yucatán or the Andes, but in East St. Louis, Illinois. In America's Southwest (in the Anasazi-Pueblo-Hopi tradition), people kept track of time by sun and moon without use of writing. Like the Aztec and Dakota, they named their months and years to correspond to the activities that took place at a particular point in the year's cycle and they carved astral representations on rock outcrops situated in special locations where they followed the course of the sun across the local horizon. Present-day Kansas once was populated with Pawnee lodges whose construction paralleled indigenous ideas of how the universe was created.

While the surviving evidence still leaves unclear just how and to what extent these cultures were related to one another, it is possible to sketch out some common denominators in the ways that North America's first astronomers regarded the sky. We can partly decipher calendar sticks and star maps made on deer and buffalo hide that name the shapes of constellations. Surviving oral tales connect many of these star patterns to questions about the origins of the world and its peoples, the chance of drought or flood, and how to propitiate the gods to assure good and avoid harm.

If you want to see earthworks in the United States that are bigger and as complex in design as Britain's Stonehenge, visit Newark, Ohio. There you will find two square miles of precisely constructed earthen octagons, squares, and circles that were built about A.D. 250 by the Hopewell people—with a modern golf course running through the middle of it. They were among the first riverside fishing and hunting-gathering communities, which had flourished since the fifth century B.C. and which organized themselves socially and economically on a complex scale. They manufactured and traded mica mirrors and awls, and collected rare materials such as copper and shell, as well as meteoritic iron from as far west as Wyoming.

But why would such a specialized, sedentary society deliberately pile up huge embankments of earth a dozen meters wide and several handspans high to make a 60-acre octagon with apertures at each point leading into the center—openings that were, furthermore, astronomically aligned? And why connect such a geometrical oddity to a perfect 321.3-meter-diameter circle by a narrow conduit. The most popular analogies that come to mind are the earth-art sculptures consisting of curtains drawn across canyons, or thousands of colored umbrellas dotting a mountain landscape—all executed by modern environmental sculptors.

One element missing from our modern analogy is people. A more reasonable hypothesis for mid-America's earthworks is that the figures were not just art for art's sake, but rather places of assembly where sky worshippers might

This aerial view of a Mississippian site in Newark, Ohio, shows earthen mounds in octagonal, square, and circular forms now interspersed between the fairways of a modern golf course. Some scholars believe that these mounds may have been dedicated to the worship of the moon, since the sight lines between the openings line up with the "standstill" position of the moonrise and moonset, which is the same orientation found in the stones of Stonehenge and Callanish.

periodically gather to talk to the gods, passing from one figure to another along avenues that still connect them together. The tops of the high banks that surround the enclosure are wide enough to accommodate a parade of individuals—leaders, perhaps, who might be looked up to for an account of what the gods held in store.

The same kind of megageometry shows up in other earthworks along the Ohio River and its tributaries. It suggests that regardless of how they built them, the people were well organized enough outside the trade and agricultural sphere to institutionalize their skywatching and probably their calendar as well.

Which sky gods were involved? Physicist Ray Hively and philosopher Robert Horn of Earlham College, in Indiana, who have surveyed the Newark site, believe that the entire assembly may have been dedicated to the worship of the moon. They discovered that sight lines between the openings of the octagon, avenue axes and diagonals of one of the squares, all mark out with remarkable accuracy the extreme, or "standstill," positions of the moonrise and moonset—reached by the lunar deity every 19 years—the very same orientations found at Stonehenge and Callanish. Studying the diameters of circles, sides of octagons and diagonals, and sides of earth squares, they were able to determine that the whole site was put together through the use of a single measuring unit 1054 feet (321.3 meters) in length.

There is no solid archaeological evidence to support hypotheses about the sacred use of the octagon/circles. Still, we get a sense that the Hopewell people were deeply concerned with geometrical and celestial harmony, and that they possessed as much skill for creating a ritual-focused environment as either the Baroque architects who built the great cathedrals of Europe or the New World builders who conceived Maya and Aztec outdoor plaza and temple complexes.

Cahokia, situated near the confluence of the Mississippi and Missouri rivers in what is now Illinois, was a large Hopewell site and an important trading center that reached its peak of development in the 13th century. At that time, it may have housed as many as 50,000 people, though most estimates put the figure between 10,000 and 20,000. The mound alignment here seems to indicate that the sun rather than the moon was the main focus of attention during the settlement's development.

The same general kind of sky symmetry is apparent at Cahokia, located near the confluence of the Mississippi and Missouri rivers. Cahokia was a significant economic and political center of grand proportions; it controlled the distribution of maize and exotic trade items over a very wide area. The axis of orientation there is cardinal, and mound alignments imply that the sun, rather than the moon, was the principal object of attention. By following the annual solar path along the horizon, rulers of this economic hub could regulate the seasonal flow of goods and services and schedule the holidays and their accompanying solar rituals that would take place when the local populace and the tributaries of the state turned out in the plaza in front of the great Monk's Mound. In a similar vein, farther down the Mississippi, French missionaries witnessed the Natchez tribes assemble in early colonial times to worship the rising sun, believed to be incarnated as their chiefs, whom they called "Great Suns."

"The Hopi orientation bears no relation to North or South, but to the points on the horizon that mark the places of sunrise and sunset at the summer and winter solstices," wrote pioneer anthropologist Alexander Stephen, who visited Tewa village in New Mexico in 1893. The map accompanying Stephen's journal suggests that unlike the Mound Builders, who arranged their own environment to make it appear that they truly lived in the sky, these descendants of the Pueblo-Anasazi of northern Arizona/New Mexico seemed to be attempting to tie their center to the world periphery by making a calendrical instrument out of the environment that surrounded them.

The Hopi were not concerned with north/south direction for their orientation or timekeeping but instead used prominent features on the horizon to delineate a celestial solar calendar. This horizon map from the 19th-century traveler Alexander Stephen shows important sun positions marked by images of the solar disk, as explained to him by native informants.

Hopi Sun Priests deposited prayer sticks, sometimes made from gourds and feathers, at shrines located along the high mesa as a gesture to welcome the sun and to encourage him on his celestial journey.

The Hopi marked the solstices, which the elders referred to as "houses" where the sun stops in his travels along the horizon. At these places along the high mesa, the priests erected small shrines. There, a Sun Priest in charge of the calendar would deposit prayer sticks, an offering to welcome the sun and to encourage him on his celestial journey. Some of these shrines have special openings that allow shafts of sunlight to penetrate particular directions, thus serving as another way to mark the appropriate time. Sometimes the Sun Priest would gesture to the sun, whirling a shield decorated with sun designs to imitate the sun's turning motion, hastening away any malevolent spirits who might impede the great luminary.

To determine the locations of shrines, which really functioned as solar foresights, the priest would need to situate himself at an appropriate sunwatching station (or backsight). These were simple stone piles giving convenient access to the expanse of the eastern horizon along which the sun would be seen to rise. One of these stations described by archaeologists around the turn of the century consisted of a flat stone with a sun face carved on top and each of the four quarters marked on its sides. The calendar specialist would sit on it and carefully sight the nearest distinct peak or valley where the sun would make its last visible slow-down prior to winter or summer standstill. Such a sighting, argues New Mexico astronomer Michael Zeilik, would anticipate the actual solstice by a few days so that the people could then have time to prepare for the ceremony. The Soyál Ceremony at winter solstice, for example, lasts nine days and its announcement is made four days in advance by the Sun Priest and the Soyál Chief, both elders of the same

127

clan. Modern studies prove that timing by the sunwatching scheme has been remarkably consistent from year to year, varying scarcely by a day or two from a seasonal mean date.

It was important to know exactly where to intercept the sun god so he could accept the people's offering; he was, after all, the one god who controlled crop growth and who knew what the future held, because he alone traveled in the underworld. The Hopi astronomer was thus vested with the important responsibility of drawing a bead on the solar deity. Failure in his task could elicit serious criticism. As one Hopi clan member explained concerning an errant Sun Chief, "He was the one responsible for all that bad weather we had last winter by being late in tracking the sun."

It is worth mentioning here that a remarkable parallel to the Hopi Sun and Mountain Watching calendar exists on the other side of the world in Siberia. Each village community of North Ossetia has a specially appointed sunwatcher who shows up each afternoon at the house of worship and watches the sunset. Like the Hopi sunwatcher, he memorizes every peak and valley in the mountain silhouette across which the sunsets progress, and then he fixes the holy days accordingly. In some villages they establish the equinox by counting the number of days halfway between the solstices and noting the sunset point at that time; in others, the calendar-keeper estimates the midpoint in space between the annual solar limits. Thus they set the year at approximately 365 days, seeing no need whatever to mark out months and weeks. Because the calendar is purely observational, no mistakes ever accumulate as they do in our leap-year method of counting time.

As I implied, carved stones and notched sticks also contain a record of the routes of the sun and moon, perhaps even those of the stars. The most well-known American petroglyph is a foot-wide spiral consisting of nine-and-a-half turns carved on a vertical cliff near the top of Fajada Butte in New Mexico's Chaco Canyon. In 1977, artist Anna Sofaer discovered that sunlight passing between a 9-foot-tall (almost 3-meter-tall) pair of upright stone slabs creates a bright dagger of light that at certain times of the year streaks across the petroglyph. The device may have been deliberately worked to produce a light-and-shadow effect. On the summer solstice just before noon, a round spot of light forms just above the spiral. It develops into a dagger-like shape that pierces the center and then slides off the bottom of the petroglyph and disappears, all within 20 minutes. Before and after the solstice the whole phenomenon appears pushed off to one side.

A second light dagger, created by another opening, falls to the side of the spiral petroglyph on a smaller design, producing a similar effect. There is little question that these early forms of Anasazi sun clocks (they may date from about A.D. 1000) were designed to function as approximate solstice—or perhaps even midday—markers. Whether they were intended to be more precise timing mechanisms is problematic, as is the theory that the sun spirals also designated the limits of the shadow cast by the 19-year oscillation of the rising full moon as part of a scheme for predicting eclipses.

Another Chaco petroglyph that has received much attention from astronomers has been proposed as a record of the great supernova that suddenly flared forth in early July 1054. We can be sure of the date, even the location of this event in the sky, for two reasons. First, other people in the world say that they witnessed it. For example, a Chinese written record, the *Annal of Sung-Hui-Yao*, tells us that a guest star appeared in the third month of Chia-yu's first year on the throne. Prior to that time, in the fifth month of the first year of the reign of Chih-ho, it had appeared in the east. They say it was as bright as Venus and that it lasted 23 days, having been visible even in the daytime. The same event was recorded in, of all places, the journal of a Christian doctor in 13th-century Cairo. By means of an astrological omen, he connected an Egyptian epidemic in that same year to the appearance of a bright star in our constellation of Taurus. The second confirmation comes from modern times. Astronomers have detected the Crab Nebula—the remains of a great celestial explosion. By studying the expansion of the supernova remnant, they have confirmed historical records about when it first blazed forth in the night sky in the constellation of Taurus.

Astronomers John Brandt and Ray Williamson have made modern calculations that place this extraordinarily bright star right next to the crescent moon on the morning of July 5, 1054, more or less the way it is portrayed on the Chaco petroglyph. Astronomers Robert Robbins and Russell Westmoreland have pointed to another possible record of Supernova 1054 on a ceramic bowl, dated on stylistic grounds to post-A.D. 1000, from the

The massive ramparts of Fajada Butte rise dramatically from the floor of New Mexico's Chaco Canyon. Carved on a vertical cliff face near the top of the butte is the spiral petroglyph seen at right. The sun's rays pass through two upright slabs of stone, creating a dagger-like shape that cuts across the center of the spiral at noon on the summer solstice. At both the spring and fall equinoxes, the dagger is off center, and an additional shaft of light lands on a smaller petroglyph beside the large one.

This Chaco petroglyph depicts an unusually bright star beside a crescent moon. It may be a recording of the great supernova that flared in the sky in July 1054. The same event was recorded in other parts of the world, where the actual date is given. A ceramic bowl, opposite, from the Mimbres Valley of southwestern New Mexico, also may be a recording of the supernova. Here the lunar rabbit, in a crescent shape, touches a star-like object with its back foot. However, a more likely interpretation of these artifacts is that they note Venus and the moon.

Mimbres Valley of southwestern New Mexico. Remember the African and Mesoamerican rabbit in the moon? Apparently the ancient Mogollon people, relatives of the Anasazi, saw it, too. The bowl shows the lunar rabbit bent into a crescent position with his back foot contacting a round disk with 23 spires, the same as the number of days of visibility of the star indicated in the Far Eastern record.

Just as other cases of light and shadow effects for timing the seasons have been proposed, so, too, there are a number of alleged representations of the great supernova event. But simpler explanations for the icons also exist. The crescent moon regularly slides by the planet Venus as morning or evening star. Critics of the supernova petroglyph theory point out that this phenomenon, a far more periodic occurrence, would more likely have attracted the attention of a sunwatcher or calendar-keeper than the unusual, cataclysmic supernova, which perhaps offers greater significance to the modern eye. One of the problems of the study of petroglyphs and carved symbols in general is that such records are open to a variety of interpretations, few of which can be supported by other material evidence.

Other North American records can be more securely linked to astronomy; for example, there is little doubt that notched sticks were used from Arizona to Canada to keep track of the months. An interesting example, made by a Winnebago chief who lived on the western shore of Lake Michigan early in the 18th century, shows an attempt to integrate lunar and solar periods. More than three feet long, the four-sided stick covers six months' worth of tally notches, one for each day. Each month is clearly set off into groups of 10 + 10 + 10 (with occasional 7, 8, 9, and 11 substitutions); these represent, respectively, the waxing, full, and waning moon intervals. Here we see the base-10 system of numeration that originated out of simple body counting employed by different people in yet another context.

Longer vertical strokes on the Winnebago stick separate the months. By analysis of the count on this and other sticks, Alexander Marshack was able to determine that to keep the 12-month, 354-day lunar year in phase with the 365-day seasonal year, the calendar-keeper had utilized intercalating 27-, 28-, and 23-day months, specially noted on another part of the stick. He seems to have inserted the appropriate one of these periods into the calendar about every three years or so. This tradition of following the moon's phases on a stick seems to have been widespread in North America and very likely dates to well before the time white people crossed the plains. But how to use such a long notched stick? For what purpose?

A 19th-century Ojibwa woman from Canada gives the answer. She records in her diary: "My father kept count of the days on a stick. He had a stick long enough to last a year and he always began a new stick in the fall. He cut a big notch for the first day of a new moon and a small notch for each of the other days. I will begin my story at the time he began a new counting stick." Then she goes on to describe how her mother began storing goods such as maple sugar and rice for the winter during the first moon phase interval.

Another medium for expressing what the ancient North Americans saw in the sky is sand. Metaphors for prescribed moral behavior, Navajo sand paintings are colorful, highly symmetrical works of art, meant to be destroyed soon after they are made. They are still used in healing ceremonies as part of the ceremonial paraphernalia for attracting a particular healing power.

Some sand paintings, analyzed as cultural maps by historian Trudy Griffin-Pierce show the constellations. These are the stars that adorn the body of a sky chanter. The stars represent old people and the paintings indicate that the elders who move around the fixed Pole Star must stay at home, close to the hearth and the central fire, near to the family in the *hooghan* where they are needed.

But what have Cassiopeia and the Big Dipper to do with old people, or with curing for that matter? The sand paintings with stars on them recreate past events in the sky; and we must remember that all the people from the Southwest U.S. to Yucatán, believe that the future is contained in the past. Past time is directly accessible to us whenever we act out a ritual. This means that when we recite a chant or make a sand painting, we afford ourselves a means of direct communication with the spirits who, for example, cause ailments in both past and future.

Every time we perform a rite we learn through the dialogue—a two-way conversation with the power spirits—how to discover and reaffirm the order in the world. "We have to relate our lives to the stars and the sun, the animals, and to all of nature or else we will go crazy, or get sick," a Navajo girl once told an anthropologist. This is the kind of organization and order a Navajo sees in sand paintings.

The fire in the middle of the Navajo hearth was brought into the hooghan of creation by Black God. A sky legend says that after earth and sky were separated, Black God walked in the door. People noticed a little cluster of seven stars called the Flint Boys that he carried on his ankle. To attract attention, he stamped his foot and the Flint Boys flew to his knee. Black God stamped once more and they jumped higher up to his hip, once

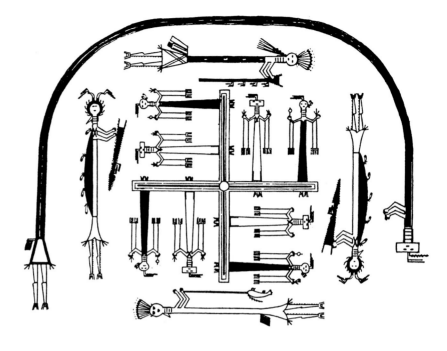

This Navajo depiction of the cosmos shows the compass cross, the swastika of sky revolution, and the gods of rivers, mountains, and rain with the rainbow god arcing above them.

According to Navajo legend, each time Black God stamped his foot a cluster of stars moved up his body until they finally settled on his forehead. This is where they remain today, visible at the top of the sky in autumn and winter, and known to us as the Pleiades. The face of the god may represent a schematic sky map with the moon crescent at the top, the sun at the bottom, and the line between them depicting the ecliptic astride which the Pleiades appear.

again and to his shoulder, and finally up to his forehead. That is where he decided they should remain forever. Because he was, by his own demonstration, clearly in charge of the sky, no one questioned him. And there they still can be seen today, high up at the top of the sky in autumn and winter—our Pleiades.

Perhaps the most extensive North American record of constellations comes from the Pawnee territory of mid-Kansas. A star map painted on buckskin (now on display in Chicago's Field Museum) was once said to have been used to wrap a sacred meteoric stone; it contains hundreds of celestial four-pointed star symbols. The Dippers' bowls (stretchers) and their handles (Medicine Man and Errand Man who follow) and Cassiopeia are all recognizable. It should not matter that these star groups are not positioned as we would place them according to our own cartographic rules. For example, Corona Borealis seems especially prominent and exaggerated at the center of the skin's surface (we will learn why shortly), and on the opposite side of the Milky Way lie the Pleiades. Farther off center Orion's Belt is positioned. A band of closely gathered tiny stars running right across the middle of the map may represent the Pathway of the Departed Spirits—the Milky Way. This intriguing buckskin map, the only surviving artifact of its kind, is reminiscent of the Aztec and Maya codices in that it was undoubtedly a sacred text of great power. But it was also a mnemonic device for remembering the many star stories that go with the appearances in the sky of the indicated constellations. All these legends would be brought out—

A Pawnee star map painted on buck-skin contains hundreds of four-pointed star symbols with many star groups recognizable, even though they are not precisely positioned. It seems probable that Pawnee holy men used this map for ceremonial rather than astronomical purposes.

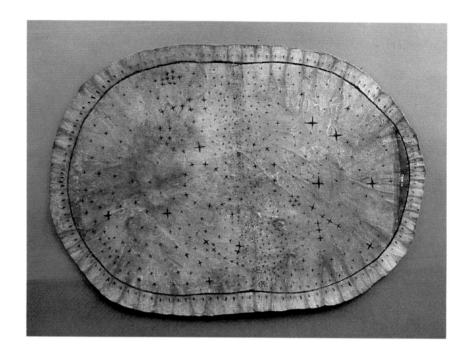

with associated ritual paraphernalia and narrative skills—at the proper time and in the designated place, as the celestial rules of order prescribed.

No matter who put them in the sky, all constellations are designed to tell stories. The Sioux are among the most colorful and fascinating North American star-storytellers. For them the stars were the very breath of the creator. According to folklorist Ronald Goodman, the Lakota constellation of the Hand (basically the bottom half of Orion) is tied to the tale told by one storyteller of a great chief who lost an arm. The wrist of the hand is Orion's Belt while the thumb is composed out of his sword. Rigel is the tip of the index finger, and Beta Eridani that of the little finger. A maiden daughter of a chief tells that she will marry only the warrior who will recover her father's arm, which was ripped out of its socket by the Thunder People. Fallen Star, a young male born of a terrestrial mother and a celestial father, rises to the task. He travels from one community to the next in the sky and earth worlds acquiring special powers as gifts from the spirits who cross his path so that he may elude and ultimately outwit the Thunder People. From them he seizes the lost arm, returns it to the chief and wins the girl.

The main purpose of this sky story is to raise questions. As young Lakota listeners look up at the Hand in the sky while they absorb the words of the fireside story, one wonders: "Why did the Thunder People take the Chief's arm in the first place?" The narrator tells them that the Chief's selfish behavior is the reason that the gods intervened and took it away. "But why is he allowed to get it back?" she asks. "Because the harmony between the gods and the peo-

133

The dome shape of the Skidi Pawnee lodge represents the structure of the heavens, and its circular plan imitates the horizon. Posts supporting the dome symbolize stars in the east. The doorway is positioned so that the sun, when rising at the equinox, illuminates the altar at the rear of lodge.

ple must be restored and this can happen only through a struggle taken on by the younger generations." Like the arm restored to the Chief's body, the regenerative power of nature must be brought back to earth each year. But this can happen only with the help of human effort. Like the freshly fertilized earth in the new year, the newborn son of Fallen Star and the Chief's daughter represents the next cycle of time. Eventually, when the Hand disappears from the sky, they say that it warns us of the people's everlasting mandate to help bring about the next cycle of life, for at this time of year all the plants begin to die.

Far from being an amusing celestial cartoon, the story of the Lakota Hand constellation—along with countless other sky stories told by native North Americans—has deep moral significance. In its disappearances and reappearances, it offers the believer a continuous affirmation of the people's role in keeping the cosmic cycles in gear and forever moving.

Another stellar classroom and medium of astronomic expression was the Skidi Pawnee lodge. The dome shape of the lodge mimics the structure of heaven and its precise arrangement and orientation, with the doorway on the east and a carefully positioned smoke hole through which to view the stars, celebrates its own built-in sky symmetry. An early visitor was told that the posts supporting the lodge are really the two Morning Stars in the east and that of the two inner poles, one is the Morning Star's messenger, the other is the Black Meteor's messenger.

Astronomer Von del Chamberlain, who has thoroughly analyzed the structure of the lodge, sees it in part as a functioning calendar. For example, he finds that the doorway is precisely positioned to admit sunlight so that it reaches an altar on the extreme west side of the interior for little more than a 20-day period spanning the equinoxes. The path of the bright shaft of noontime sunlight entering the smoke hole would change with the course of the year, coming only part way down the wall at winter solstice and all the way down to the floor by mid-February. The points where the sun touches the boundary between wall and floor could be mentally segmented out to depict different times of the year. By the time the inhabitants abandoned their winter lodges for their outdoor *tipis*, the solar image would have migrated to the position closest to the center of the lodge. In addition, the star groups recognizable from Pawnee lore also could have been sighted in the house apertures. For example, the Pleiades can be first glimpsed briefly through the smoke hole by an observer sitting at the wall along the lodge's axis of symmetry just before sunrise in late July, then again just after sunset around the time of the winter solstice. The Chiefs in Council (Corona Borealis) would enter the smoke hole in direct opposition to the times for the Pleiades. This may explain their opposed location in space on the Pawnees' buckskin star map, according to Chamberlain.

Can we say that the Pawnee lodge was an astronomical observatory and that its entryway and smoke hole were intended solely as calendrical time slots? Chamberlain thinks it more likely that these shelters from the cold winter air were living astronomical schoolrooms through which visual illustrative star and sun scenes were used to enhance the real-life, "just-so" stories and moral tales told within their warm confines.

There is no question that the occupants made observations through the sky-house apertures: "Observations were also taken through the smoke hole of a lodge, by taking a seat west of the fire at sunset and noting what stars could be seen," wrote ethnologist James Murie in *Ceremonies of the Pawnee*, his book on life among the Pawnee at the turn of the century—though he gave frustratingly few details. So, for the Pawnee celestial skywatcher, then, a sky dome need not be as precisely constructed as one of our modern observatories or planetaria.

Whether the mode of expression be a house of many apertures, an impermanent design composed of grains of sand, or a more conventional stick with notches, our brief excursion across the North American continent reveals that though the sky is pretty much the same across the length and breadth of the land, the text on which America's first astronomers described that sky is unimaginably diverse.

The city of **Cuzco** was designed to meet the Inca need for order. Originally, it was laid out in radial segments to enable each particular the proper place and at the appropriate time in the calendar year.

social class to perform its assigned function in

9

PATHWAYS TO THE STARS: SKYWATCHING IN THE ANDES

Would we expect the Inca of Peru to have developed a sophisticated system of astronomy like the Maya? In the first place, they did not produce any sort of written record as we know it. And second, we tend to regard them as New World Romans—too occupied with war and conquest to turn to matters as esoteric as star-gazing. But the establishment of an empire stretching over nearly 30 degrees of geographic latitude gave them reason enough to create a calendar based on precisely timed celestial events, which they employed in regulating civil, agricultural, and religious dates on a national basis. They encoded their calendar in architecture and through a notation system of their own that is quite different from ours.

Of all the reasons we can give for the success of the Inca empire, those we must take most seriously are the strict order and the high degree of organization that was built into every component of it. Said an early Spanish visitor to the highland Inca capital: "Nowhere in the kingdom of Peru was there a city with the air of nobility that Cuzco possessed…compared with it, the other provinces of the Indies are mere settlements. Such towns lack design, order, or polity to command them, but Cuzco has distinction to a degree that those who founded it must have been people of great wealth."

Cuzco is situated in latitude 13½°S at the junction of two rivers in a 10,500-foot-high (3200-meter-high) mountain valley in the central Andes. The Inca may have called it Tahuantinsuyu, or "the Four Quarters of the Universe," because of its basic quadripartite plan. The city is segmented into halves called upper and lower Cuzco. Each half in turn is split into two sectors, or *suyus*. Now, none of these four regions occupies a 90-degree segment of a circle, for the principal rationale for dividing the city this way had more to do with the watershed environment and kinship rank than considerations of pure geometry. The lines between suyus demarcate the flow of underground water in the Cuzco valley, which naturally follows a radial plan in this montane environment. Suyus were intended to serve as an organizing principle to define water rights for the kin groups who farmed the wedge-shaped plots of land between the river valleys. The people believed that the reception of the underground water was part of their birthright, and that it came directly from their ancestors who resided in the body of the earth-mother (Pachamama), a deity whom they honored and nurtured by offering sacrifices to feed her.

Four major roads departed Cuzco—one from each corner of the central square. These served as the dividing lines among the suyus. Theoretically they extended to the remotest domains of the Inca empire—as far as Quito to the north and central Chile to the south. Suyus were ranked, as were the hierarchically organized kinship groups living within them. The organizing principle was the moiety, or two-part, division, and it was based on whether they were located up-river in higher ranking Hanan or down-river in lower ranking Hurin (Cuzco).

But what do kinship and geography have to do with the stars? The *ceque* system was a giant cosmogram—a mnemonic map built into Cuzco's natural and man-made topography. It served to unify Inca ideas about religion, social organization, calendar, astronomy, and hydrology. The 17th-century Spanish chronicler Bernabé Cobo has left us a thorough and detailed description of this rather abstract system. It appears to have consisted of a number of imaginary radial lines (ceques), grouped like spokes on a wheel according to their location within each of the four suyus. The wheel's hub was the Coricancha (called by the Spaniards the Temple of the Sun). Covered over today by the Church of

At its peak toward the end of the 15th century, the Inca mountain empire stretched along the western edge of South America through 30 degrees of latitude and included several great cities, among them the southern capital of Cuzco and the northern one, Quito. Nazca, near the coast, was the site of an earlier culture.

Santo Domingo, this structure was the most important temple of ancestor worship. The Coricancha was related as much to the underworld as it was to the heavens, for it is within the earth that all our predecessors live. Cobo lists nine ceques associated with each of the suyus: the northeast (centered around the district named Chinchaysuyu), southeast (Antisuyu), and southwest (Collasuyu), while 14 ceques were associated with the northwest (Cuntisuyu) quadrant, thus making a total of 41 radial lines.

According to Cobo, each ceque was assigned one of a set of three hierarchical groups that represented the social classes that tended to it and maintained shrines called *huacas* (sacred places) located along it. There were ceques said to be cared for and worshipped by the primary kin of the Inca ruler. Following anthropologist R. Tom Zuidema's description of the system, let us call this sort of ceque type (a), those that were worshipped by his subsidiary kin (b), and ceques tended by that segment of the common population not related to the ruler (c).

Cobo's tract implies that the assignments on the hierarchy of worship rotated sequentially (a),(b),(c); (a),(b),(c); etc., as one proceeded from one ceque to the next within each suyu. This ranking system followed still another rule of order. It went all the way around the horizon in a clockwise direction in the northern suyus and counterclockwise in the south.

According to Cobo, each ceque was traceable by the line of huacas leading outward from the Coricancha across the landscape. At these huacas, worshippers communed with the gods who controlled the cosmic forces. They left offerings at the openings in the body of Pachamama. The placement of the huacas must have been rather important, for Cobo goes through the considerable trouble of locating and describing each one in detail. There were 328 in all; some were natural rock or man-made temples, intricately carved rock formations, bends in rivers, fields, hills, springs or other natural wells called *puquios*, and even impermanent objects such as trees.

Read Cobo's careful, typical description of two of the 328 huacas that make up the system of ceque lines:

> Chinchaysuyu—ceque 6, huaca 9: A hill called Quiangalla that is on the road to Yucay where there were two monuments or pillars that they had for signs and when the sun arrived there it was the beginning of summer.
> Cuntisuyu—ceque 13, huaca 3: Chinchincalla is a large hill where there were two monuments at which, when the sun arrived, it was time to sow.

Structures such as the Fortress of Sacsahuamán at Cuzco indicate why the Inca were sometimes called the Romans of the New World, but it is still not known exactly how they built such massive structures with only simple tools and without wheels or draft animals. Though the huge stones weigh as much as 200 tons, they fit together so well that they have withstood Andean earthquakes unscathed.

Invisible radial ceque lines, recognizable in the mind's eye because they link sacred places of worship, emanate from the center of Cuzco. The lines extend out to the distant mountain horizon, often connecting with bends in rivers where offerings could be placed. This idealized schematic map is only an approximation of a much more complicated system.

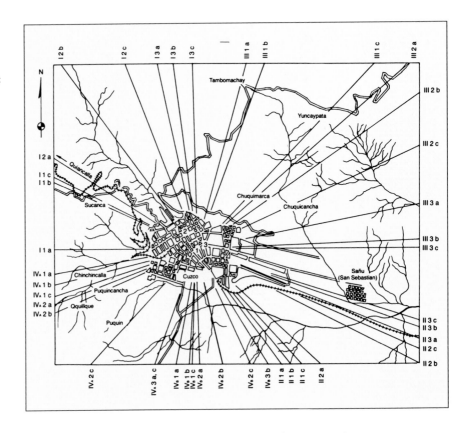

Other huacas of this ceque system also served purposes that were astronomical. For example:

> The seventh (huaca of the 8th ceque of the Chinchaysuyu quadrant of the city) was called Sucanca. It was a hill by way of which the water channel from Chinchero comes. On it, there were two towers as an indication that when the sun arrived there, they had to begin to plant the maize. The sacrifice which was made there was directed to the sun, asking him to arrive there on that hill at the time which would be appropriate to planting, and they sacrificed to him sheep, clothing, and miniature lambs of gold and silver.

Note the unusual detail in this passage—especially the rather concrete information that connects the flow of water to sky events. Cobo clearly ties sky knowledge to aspects of everyday life. An anonymous chronicler describes the passage of the sun by four pillars on a hill called Cerro Picchu, which overlooks Cuzco from the northwest. His description resembles the huaca called Sucanca that is described in the previous passage, even though Cobo mentions only two towers there:

> When the sun passed the first pillar they prepared themselves for planting in the higher altitudes, as ripening takes longer.

> When the sun entered the space between the two pillars in the middle it became the general time to plant in Cuzco; this was always the month of August.

> And when the sun stood fitting in the middle between the two pillars, they had another pillar in the middle of the plaza, a pillar of well worked stone about one estado [6 feet, or almost 2 meters] high, called the Ushnu, from which they viewed it. This was the general time to plant in the valleys of Cuzco and surrounding it.

The observer in charge of the calendar must have stood somewhere in the present-day main plaza of Cuzco, the Plaza de Armas, a few hundred meters from the Coricancha, to watch the sun set. But the logic of this Inca space-time calendar is not as simple as it might appear at first glance. If we take the chronicler at his word, the ceque itself could not have been the sight line, nor the center of the ceque system the point from which the observations were made. The data simply do not fit. More likely, the sighting station was the Ushnu, called "a fountain of well-worked stone one estado [about 6 feet, or 2 meters] high" that once stood in the plaza. Evidently the Inca used the pillars to chart the time for planting in different elevations in the vertical environment of Cuzco, marking out the day-to-day horizontal course of the sun across the row of towers.

The time of year when the sun passed the pillars was, according to the chronicler, the middle of August. This is exactly opposite the time of year when the sun crosses the zenith. Recall how important the zenith—the overhead point in the sky—looms in most tropical systems of astronomy. The Inca discovery that these important agricultural dates coincided with one of the prominent, conveniently visible celestial phenomena in the environment of Cuzco, may have led them to regard the act of planting and the passing of the sun underneath the world as like-in-kind events. At this time of year, an Inca legend says, the earth mother "opens up." Pachamama is then at her peak of fertility and can be penetrated both by the tiller (with his plow) and the sun (with his rays). The sight line across the landscape that connects the rising and setting sun on the days of overhead and underfoot passage is an Inca way of expressing in horizontal space the vertical nature of the ecology of the Andes. In other words, as the sun advances horizontally the time to plant advances vertically.

Cuzco's ceque map was not just a directional scheme incorporating significant astronomical events that happen at the horizon. It was also a seasonal calendar, with each huaca representing a day in the year, and clusters of ceques signifying the lunar months. There are many calendrical numbers hidden away in the ceque system. According to Zuidema, the number of ceque lines—41—doubled served as a count of three sidereal lunar months (the time it takes the moon to return to the same station of the zodiac regardless of its phase). He has also noted that the time the Pleiades are absent from the sky corresponds to the 37-day difference between the seasonal year of 365 days and the year of 328 days counted by the huacas.

The association of planting, irrigation, and the Pleiades star group probably developed when people recognized that the period of absence of the prominent star group from the sky coincided with the time between the end of the harvest and the beginning of the next planting season. It became leftover—or uncounted—time in the annual calendar, similar to our 12 days of Christmas or the five bad-luck days at the end of the Mayan year. There is evidence that the Inca used an alignment of the walls of the Coricancha itself to correlate the place where the Pleiades rose with the timing of their major festival of Inti Raymi during the June solstice. In 15th-century Cuzco, the Pleiades returned to view just before the solstice, having been blocked out by the light of the sun for the approximately 37-day period. Interestingly, the terminal huaca of the ceque that indicates this direction is called Susumarca, which is one of their names for the Pleiades.

For convenience, the Inca slightly adjusted their monthly intervals so they would coincide with significant periods in the agricultural cycle, such as the time of plowing, planting, the appearance of water, and harvesting. There was much concern that certain feasts and rituals be celebrated during a period commenced by particular named full moons—like our antiquated habit of beginning the harvest during the month initiated by the appearance of the Harvest Moon or of opening hunting season with the Hunter's Moon. The 12 months of the lunar synodic calendar were cleverly subdivided among the population of Cuzco and vicinity in such a way that each social group was assigned the responsibility to perform the ceremonies and sacrifices at those huacas associated with the group's particular month of the year in the calendar-counting scheme. Like the Aztecs' mandate to feed the sun in order to keep it on course or the Hopi sunwatchers' duty to encourage the luminary on his way, the kin groups who made up Cuzco's populace had a role to play in articulating the Andean pathway of time.

As we have seen, the city of Cuzco was a picture of order in its very structure. It was in essence a map of itself, one that unified, bounded, and subdivided natural and social space and time. The natural order that the Inca and their subjects perceived in the landscape served as a means of structuring social order. The whole system worked when each particular class performed its assigned function in the proper place along its ceque line at the correct time in the calendar. In this confrontation between nature and culture in a harsh and variable agricultural environment, the ceque system emerged as a clear scheme the Inca royalty devised to prescribe proper human action. It was a brilliant idea based upon residence and kinship in a radial, four-fold geographic framework. Astronomy was a part of that order, but it was inseparable from all the other components. The lesson is that it makes no sense to pull astronomy out of context.

The Inca *quipu* (at right and in use below by a *quipu camayoc*—literally an information-keeper—in a 17th-century illustration) generally consisted of colored strings attached to a main cord, and served as part of a counting system. Knots on each string represented the numerical equivalent of the items being counted. Colors seem to have designated different data or events. For instance, black is assumed to have represented time, and red war or warriors. Some also see the ceque system as a giant quipu overlying *Cuzco*; the cords become ceque lines, and the knots are the huacas.

On the desert coast of Peru the famous Nazca lines may bear ancient traces of the ceque system. Built by the culture of the same name in the first few centuries A.D., the lines consist of geometrical figures and animals that bear a distinct relation to those painted on the famed Nazca pottery. But the overwhelming number of figures, etched on the dry desert by the process of removing the oxide-varnished topsoil, consist of straight lines. There are some 800 of them, some several miles long, emanating from 62 focal points. The whole organized pattern resembles an assemblage of ceque systems tied together in a network spread across 100 square miles of the elevated flat surface situated between the Ingenio and Nazca rivers. Locals call it the pampa.

Our studies on the pampa floor leave little doubt that the lines are neither artwork to be viewed from above, nor runways for extraterrestrial aircraft as some popular publications suggest. Rather they were pathways, probably walked over by participants in a rain-bringing ritual. Most of the lines begin and end at water sources. But sun-watching may have played a role, too. A significant number of Nazca's straight lines point to the place where the sun rises during the season of the year (early November or spring in the Southern Hemisphere) when water begins to run in the rivers and underground canals that dot the edge of the pampa. It rarely rains in Nazca; therefore one can anticipate the arrival of the life-giving liquid only by watching dark clouds form in late afternoon over the distant Andes or by following the horizon-bound course of the sun.

What about the curious two dozen-odd animal and plant figures? Are they constellations? Maria Reiche, a former teacher of mathematics who has worked in Nazca for most of her life, clings to that theory. She detects the fig-

Measuring several meters in length and drawn with one continuous line, this design of a hummingbird is one of the most fascinating examples of the Nazca lines. Representations of other animals, including fish, monkey, and lizard, abound on the pampa.

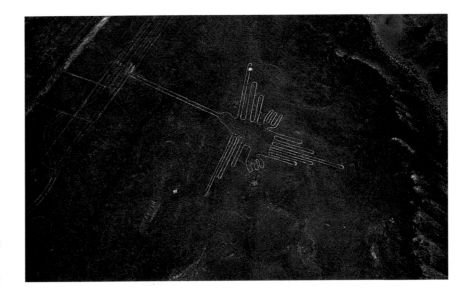

The giant markings found scattered across the Nazca desert seem to be pathways laid out and used by people in rain-bringing rituals. Most lines begin and end at a water source. Sunwatching also may have played a part in the positioning of these patterns, since many of the lines point to the spot where the sun rises when water begins to run in the canals.

ure of our Orion in the giant spider, and a huge drawing of a monkey with a labyrinthine tail becomes Ursa Major. But aside from such superficial and subjective resemblances, there is little other evidence to suggest that the great ground effigies on Peru's pampa convey any more astronomy.

Anthropologist Gary Urton has traced a number of concepts and myths among living Quechua-speaking people—descendants of the Inca—that connect all the way back to Inca times, if not earlier. He found, for example, that the Pleiades are still called "collca," or the agricultural storehouse, and that Alpha and Beta Centauri—among the few bright stars present in alignments tied to huacas of Cuzco's ceque system—represent the "Eyes of the Llama." The latter is one of a parade of dark-cloud constellations that, along with star-to-star constellations like our own, comprise the Milky Way, which is so much more prominent in the Southern than in the Northern Hemisphere. Other animals in the zodiac-like parade across the firmament include fox, partridge, toad, and a great anaconda. Urton argues that the Milky Way was a sort of functioning environmental calendar—like ancient Cuzco's sun pillars. Thus the interval during which many of these sky creatures were visible corresponded to periods when their terrestrial counterparts were active. For example, the partridge-like bird called Yutu makes its heliacal rise early in September and disappears from view in mid-April. A chronicler tells us that the beginning of this period is just the time farmers needed to guard their crops against these birds. Likewise, real toads re-emerge from the earth just about the time the toad constellation first clears the horizon, bringing the rain with him. When the Spanish chronicler Polo de Ondegardo wrote that "all animals and birds on the earth had their likeness in the sky in whose responsibility was their procreation and augmentation," one has to wonder whether he realized that he was dealing with a highly organized astronomical system of tracking environmental time. More than a memory assist, these star groups held the power within them to bring good or bad fortune to the hard-bitten farmer of the high Andes.

The idea that the sky is a blueprint for everything beneath it also flourished on the other side of the Andes in the rain forests of the Amazon and the Orinoco. The Desana of Colombia represent one of the few thorough case studies. These people divide the year into two rainy and two dry seasons and reckon the central point of their calendar by the place where the shaman's staff will cast no shadow when held upright. Like a horizontal lid, a celestial hexagonal template consisting of the bright stars Procyon, Pollux, Capella, Canopus, Achernar, and one of the stars of our constellation Eridanus (Tau 3) overlies the earth at sunrise/sunset, just as the sun is positioned at the equinoxes. At this time, when heavenly symmetry is in force, a vertical shaft of sunlight is said to fall on a mirror-like lake below, thus fertilizing the earth. Furthermore, the original tribes were said to be six in number and they still organize themselves socially in a hexagonal model.

Anthropologist Gerardo Reichel-Dolmatoff believes the canopy crystal also served as an architectural model, for the Desana longhouses are built on a six-sided plan. Informants told him that each vertex of the hexagon consists of a housepost that can be identified with one of the basic support stars. The bisector of the hexagon on earth is a ridge-pole identified with the Pleiades, which rise in this area today just after spring equinox, thus signaling the beginning of the main fruiting season. To commemorate the event, piles of palm fruits identified with the Pleiades are heaped at the center of the house. Orion is another symbol of the center, no doubt because he lies in the middle of the astro-hexagon and because his belt perfectly straddles the equator. Once again he is a hunter, but unlike his Old World counterpart, Orion of the jungle is far more ambivalent. He possesses many incongruent character traits, such as ancestor and hero on the positive side, sinner and victim on the negative. For example, he is frequently portrayed as an adulterer, an image that has been documented in places as far away as the land of the Arawak in Guyana. There the one-legged rascal even seduces his own mother-in-law! Among the Bororo of Brazil the box-shaped pattern of the lower half of Orion is one of a number of turtle constellations (Corvus and Scorpio are others). In their world, the order of nightly and seasonal time is marked by a parade of celestial turtles, just as the many species of amphibian in the terrestrial environment below gave signals to hunter or fisherman regarding the "catch of the day."

The Desana mark out youth, maturity, and old age on their sky hexagon and bind it to their terrestrial six-sided domicile. Expressed in a dance, a symbolic journey around the longhouse typifies the cyclic journey of both men and women through life. Each of the stellar vertices represents a significant mark-

Anthropologists believe that the heavenly canopy served as an architectural model in many cultures. The traditional Yekuana wattle-and-daub roundhouse is similar to the Pawnee lodge in that it features a roof window that receives the sun. The path of the sun on the interior wall acts as a calendar.

ing post along life's road. Men, for example, move clockwise through Capella (naming) to Pollux (initiation) to Sirius (marriage). Women travel counterclockwise, but only until they arrive at Sirius; then they turn about and join their husbands. When all return to their starting point, the Pleiades and Aldebaran, they are reborn, precisely on the equinox line.

We encounter cosmic houses among both the Plains Indians of the United States and indigenous tribes of the ancient Hawaiian Islands. In South America we find yet another celestial domicile—the Yekuana household—that mirrors the astronomical model.

The Yekuana Indians inhabit the headwater regions of the Orinoco and a number of its tributaries in Venezuela. They build large communal roundhouses consisting of cylindrical outer walls of wattle-and-daub and conical roofs of thatch. The blueprint was given them by Wanadi, the Sun, in the form of a model house the solar god built for himself on a mountaintop. Reminiscent of the Pawnee lodge, a single roof window is eccentrically placed to receive the sunlight during the dry season and to permit passage of smoke out of the living quarters. The window also serves the function of making out the yearly calendar by following the sunbeam—entering the window—sweep across the interior opposite wall. Daily time can be measured by the simultaneous ascension of the light shaft up the central zenith pole of the house. The light beam also serves the religious purpose of following the sun upward through the interior roof space, symbolizing the cosmic vault, to the apex of the structure where the Supreme Being is at home and where he plays out the game of the seasonal cycle.

Control of the seasons is also in the charge of the heavenly deities among the Barasana who live farther upstream from the Yekuana. Anthropologist Stephen Hugh-Jones observes that the Barasana divide their zodiac into two halves. There is an Old Path centered on our Scorpio but all the constellations are very different from our own. Old Path consists of all things associated with death and decay: poisonous snake and spider, the headless corpse of an eagle, a vulture and, oddly enough, the corpse of a woman stung to death by wasps. In stark contrast, the New Path contains only good things—things of subsistence like a fruit arbor, fish-smoking rack, Jacundá fish, big and little otter, and cray-fish. The head constellations of each half control that portion of the season, the Pleiades or Star Woman marking the beginning of the good half, while the Caterpillar Jaguar, a jaguar with a snake for a tail (the equivalent of our Scorpio), connotes the Old Path.

In this imaginative cosmology, Star Woman is in charge of the seasons and agriculture. She appears at dusk on the eastern horizon to mark the end of the rain and the beginning of planting and she disappears in the west after sunset in April to end the dry season and commence the heavy rain that fertilizes the manioc. But Caterpillar Jaguar does just the opposite, dominating the skies of the April-November rainy season. Scarcity of food and an increase in disease are attached to his presence. But fortunately, they say that he also brings the Caterpillar People, a species of moth that pupates during the rainy season when it falls down from the trees above to become a staple in the local diet during these otherwise hard times. Men still dress up in bright feather patterns and painted body stripes in imitation of the caterpillars. They dance, as they say, to celebrate the movement of the seasons for which they believe them-selves responsible.

Notice that so many of these rain forest astronomical observations seem to focus on the ascent and descent of constellations across the heavens, their over-head positions signaling the zenith of the worldly activity with which they are associated. This habit contrasts with the horizon timings of those indigenous astronomers who, like the Inca or the Hopi, live in highland settings where high-ly visible distant mountains figure in the cosmological picture. Despite the differ-ences, the overriding similarity in all these astronomical cycles is that they operate as a mechanism for the understanding and expression of the cycle of human sub-sistence. Yet, even though every culture ultimately must run on its stomach, the sky concepts we have recounted in this chapter also orient themselves to even deeper thoughts that must have been brooding in the minds of those who devised them. The sky raises questions about other people's concepts of the real world—where the forces of nature originate, who controls them, where the human condition fits into the picture. Our examples, many of them from living cultures, should make it very clear that sky timings need not be concerned solely with practical matters, but also that the stars can be used to create colorful metaphors that express meaning in the deepest religious and philosophical sense.

These Polynesian seafarers navigate their outrigger through Pacific waters in much the same kind of vessel as their forebears used. The skills of the differences in ocean swells—once passed down through the generations—barely survive today.

navigating by the stars and by an awareness of

THE VOYAGING STARS OF OCEANIA

Until the 20th century, a navigator of the mid-Pacific islands of Oceania would have been as respectable a profession as a neurosurgeon, expert trial lawyer, or perhaps a corporate executive in our culture today. Knowing how to get from one place to another in an environment consisting almost totally of water—an ocean with wind, swells, and currents—requires carefully cultivated skills. The experienced navigator was also a perspicacious astronomer who needed to locate and memorize all 36

The vast expanse of Oceania, with its myriad islands scattered across the world's largest ocean, reaffirms the need for accurate navigational techniques for the people who sailed this area.

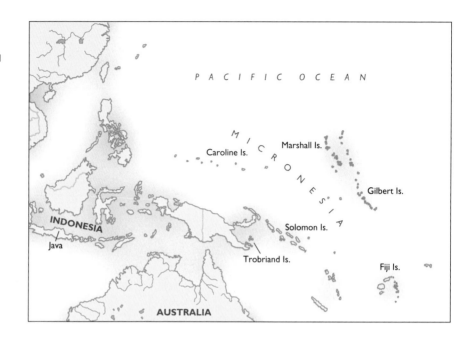

linear star-to-star constellations on an imaginary three-dimensional compass, and who needed to be aware of which island bearing was matched by a particular star group that appeared over it. Some seafarers of Oceania learned to use the stick chart, a combination map and speed indicator, for on the surface of the moving water it is more difficult to judge travel time than on land. Training in a "stone canoe"—like the handful we still find dotting the shores of otherwise deserted coral atolls—would also have been an indispensable part of a seafaring astronomer's education. Like a modern flight simulator, these fixed stone structures align with the constellations to indicate directional bearings. Once memorized, these directions can be followed in real seagoing craft that would take the learned navigators to islands of destination far beyond the horizon.

But while the astronomy of Polynesia, Micronesia, and the maritime environment of the Southeast Asian Archipelago has its practical aspects, star lore is as filled with the same creative imagination as any we find in the ancient Aegean, tropical South or Mesoamerica. For example, there is the simple story about the threat to the crops. Thieving Venus pops up as evening star and taps palm wine from the trees, which are carefully tended by the hard-working vintners. The burglar from the sky can be subverted only by another kind of celestial spell that restores the crop to normal. Here, in the midst of a vast ocean, we discover perceptive people cleverly using sky images to devise vivid metaphors that became the thread from which they wove the fabric of their extraordinarily rich mythologies. And they did it all under a very different sky dome from the one you and I perceive.

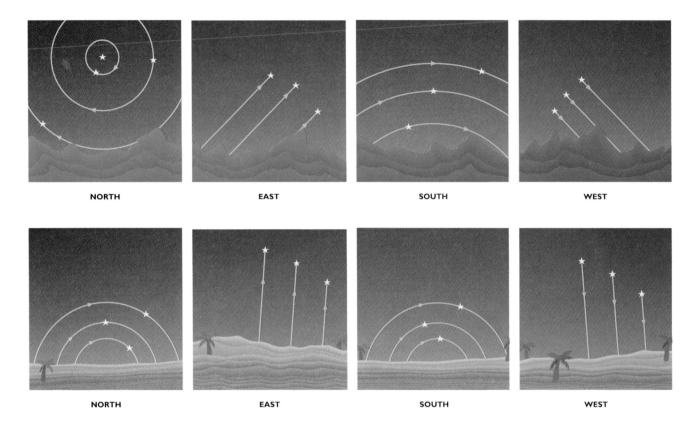

NORTH   EAST   SOUTH   WEST

NORTH   EAST   SOUTH   WEST

The movement of the stars appears quite different depending on the location of the observer. The top series shows the sky as seen by viewers who look in various directions in temperate northern latitudes. Here stars in the east and west glide along an angled path to the horizon. The lower series shows the view that ancient observers, such as Polynesian and Mayan peoples, had from home latitudes much closer to the equator. Note that for them the stars in the east and west move in an almost vertical direction.

In the illustrations above, you face west after sunset in two different locations. Your position determines the angle at which you see things streak across the sky. On the bottom, you are positioned in the tropics near the earth's equator, while the top views show the sky as seen from a relatively high northern latitude. Mayan and Polynesian skywatchers would have seen the view portrayed in the bottom illustrations. For them, all the stars plunge straight downward in the west (and rise straight up in the east). But our Western Greek and Babylonian ancestors, who lived much farther from the equator, would have watched the stars glide along more horizontal nightly paths. Those who witnessed the night sky from the relatively high latitude of Stonehenge saw an even more exaggerated horizontal motion. Compare what a tropical observer would see looking northward with what people in high latitudes would observe. Note that the people in northern latitudes, whether ancient Chinese or Anasazi, perceive all motion pivoted about a point high in the sky, while their southern latitude counterparts from Central Africa or ancient Peru barely see the Pole Star, which in these latitudes is positioned too low in the sky. For the tropical observer, the scene looking southward is about the same as the northern view; for viewers in the temperate latitudes, however, the south and north sky views are highly asymmetric.

Carefully positioned stone slabs, called "stone islands" or "canoes," are used to teach navigational skills. A pupil, seated at night between the stones, faces in one of the cardinal directions and memorizes the constellation closest to his desired destination. In daylight the stone model becomes an "island" and is used to illustrate wave lore with the triangular corner stones representing wave formations.

To summarize, those who live in the tropics witness the same motion of astronomical bodies in the day and night sky—straight up in the east, straight down in the west. The observer seems to be at the center of things with symmetric northern and southern hemispheres of the sky behaving identically. By contrast, the observer outside the tropics sees the center of all sky motion as a fixed point eccentrically pivoted high above his head.

The tropical principle was applied by navigators on a small Pacific atoll in the Gilbert Islands. Its northern shore is dotted with half a dozen pairs of parallel rough-cut slabs, each about the size of a person. They are arranged horizontally and cemented into the ground. One pair points to the neighboring island of Tamana 50 miles (80 kilometers) distant, another to Beru Island 85 miles (140 km) away, while a third gives the alignment of distant Banaba, 450 miles (700 km) over the horizon. Islanders called them "stone-canoes," or "the stones for voyaging." Locals say the slabs were used as instructional models by their ancestors to set directions for inter-island navigation. Each pair of stones aligns with the place where certain stars appear or disappear on the sea horizon at different times during the night. For example, in August the bright star Regulus aligns with the Tamana stone at sunset, while at midnight Arcturus gives the same bearing. The navigator simply memorizes a "constellation" that consists of a long vertical chain of stars associated with the island he wishes to visit; then he steers the canoe toward them. In effect, the star-rise positions become the points on a star memory compass whose elements are transmitted by oral tradition. Such correlations for accurate long-distance sailing must have required many generations of trial and error to develop.

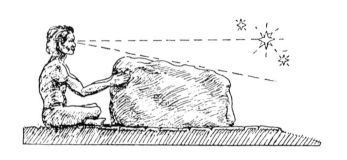

It is important to stress that linear constellations and a star compass are mental constructs and possess no analog in the cosmology and astronomy of the civilizations of the classical world. Both were prompted by environmental considerations related to ocean and sky. In other words, the orientation of the sky in the tropics made accurate ocean navigation possible by this method.

Such techniques cleverly circumvent the conventional magnetic compass and other astronomical contrivances of our own culture, but we in the far north (or south) could never make use of them. Consider the problems of a North Pacific or South Atlantic sailor who might try to adopt the system used in the Gilbert Islands. As soon as his guide star appeared over the waves, it would

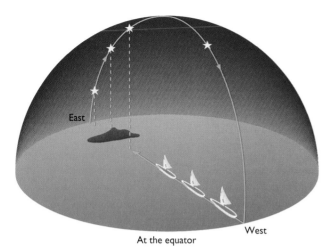

At the equator

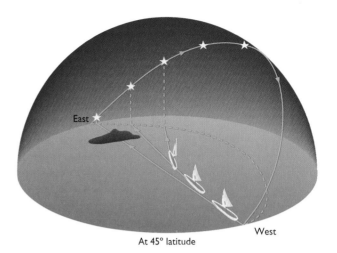

At 45° latitude

The difference in navigation techniques mitigated by nature between equatorial and more northerly or southerly latitudes is shown here. Taking a "fix" on a star will lead a sailor in his desired direction if he is close to the equator (*left*), but in higher latitudes (*right*), his course would deviate considerably as his star moved laterally across the sky.

immediately begin to move laterally, relative to the direction where it was initially sighted. Unless a substitute star were immediately available, the seafarer's course would deviate radically from a straight line. Sailing from New England to Great Britain, a ship moving at 30 knots would be thrown off course 10° or 5 miles in the space of an hour.

Sailing through the South Pacific in the mid-18th century, Captain Cook had spoken of the way native navigators used bird-flight directions, swells and wind, sun and stars to get about, but he never fully realized the extent to which the people he encountered had developed the art of celestial navigation. Much later, in 1890, a Tahitian navigator recited these instructions to a visitor: "If you sail for Kahiki (Hawaii), you will discern new constellations and strange stars over the deep ocean. When you arrive at the Piko-o-Wakea (equator), you will lose sight of Hokupua (the Pole Star) and then Newe will be the guiding star and the constellation of Humu will stand as a guide above you." Unfortunately the names of the celestial guides in our language have been lost.

Back in the 1940s astronomer Maud Makemson took an interest in the old Hawaiian skywatchers. She noted that, at latitude 20°N, they made practically no reference to Western inventions like the ecliptic and the celestial equator, but like most tropical people, they accorded great importance to the cardinal points of the horizon. Like the Maya, they believed pillars in each of the four directions supported the sky. They named these pillars as if referred to the sun situated at the east point of the horizon (*alihilani*) facing along an east-to-west axis. North (*kukulu akau*) was the right-hand pillar (also meaning the direction "up"), and south (*kukulu hema*) was the left-hand pillar or "down." The observer's position (*piko*) lay under the zenith point (*hikialoalo*), and the rising and setting points of celestial objects were called *hiki* and *kau*, respectively.

The Gilbert Islanders, who lived in latitude 3°S, segmented the sky into named zones formed by slicing the celestial hemisphere several times in the vertical direction (parallel to the east-west line) and by cutting it with another set of lines running parallel to the horizon. The vertical lines were the ridge poles (the great circle of the meridian) and rafters (small circles) of the great cosmic house we live in, while the horizontal arcs of the celestial sphere signified the crossbeams that supported the rafters. In this sky-house analogy, which we have encountered so many times before, the location of a star could be described in terms of its position in one of a collection of imaginary boxes that divide up the sky.

Unlike our cosmological models, which consider neither earth, sun, nor even our Milky Way Galaxy to be at the center of the universe, the sky dome the Gilbertese imagine, and can witness ever-present in the frame of their own family domicile, is really a multi-layered set of hemispheres centered on the home chiefdom.

Another way to get around in mid-Pacific was to use the stars regularly recognized to pass the overhead position as seen from different islands. Theoretically, we know that the declination (the angular distance north or south of the celestial equator) of a star that passes the zenith is the same as the latitude of the observer; thus, every island could be associated with one or more of its own zenithal guide stars depending on its latitude. Put in our terms, a navigator would have reached the parallel of geographic latitude of his destination when the arc of a guide star's course crossed the overhead position. Thus, Sirius is the guide star for the Fiji Islands (latitude 17°S), Rigel, in the constellation of Orion, that of the Solomon Islands (latitude 7°S), and Altair that of the Caroline Islands (latitude 9°N).

Navigation in longitude, however, is a different story. As we know it, this term was not conceptualized by indigenous sailors, there being no method known—nor even a desire—for the islanders to deal with this task by celestial observations alone. Although Old World navigators found it necessary to use timekeeping mechanisms from sand clocks to chronometers in order to set sail in the east-west direction, the people of Oceania employed a knowledge of wind and oceanic currents combined with astronomical observation; that is, they relied on purely natural forces to circumvent the problem. That they managed to get about quite successfully proves that a knowledge of longitude and latitude is not a necessary concept for all skilled navigators.

You would understand the necessity of the peoples of Oceania to rely upon the sky dome if you have ever traveled the ocean in a ship. But you might imagine it even better by making the trip in a canoe. Imagine gliding along a vast ocean at the level of the rapidly moving water, a fixed sky being your solitary reference frame. No mountains with notches, no man-made structures, neither sighting posts nor permanent pathways guide your way, only the symmetric sky above. Day after day, as you slip past distant islands on the horizon, you might begin to wonder who, or what, is really moving. Perhaps your canoe is the real center of the universe and the islands, all reckoned by the stars, the

A network of straight and curved reeds bound together form a "stick chart," a device used by the Marshall Islanders as an aid to memorizing wave and swell patterns that occur between their islands. The central, vertical stick represents the course to be followed and the bent reeds on each side signify the eastward- and westward-moving swells.

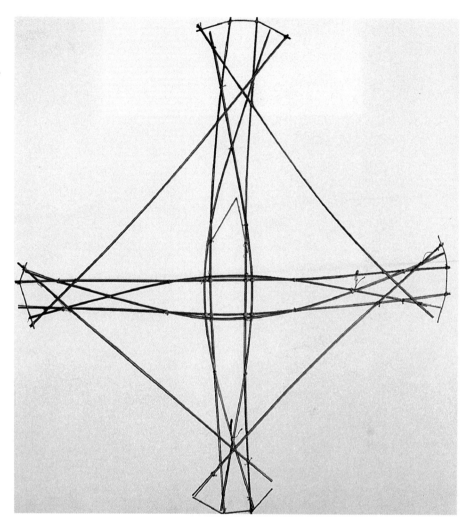

moving objects. Such relativistic perceptions are the source of the system of *etak*, a scheme used by Puluwatese navigators to track distance interlinked with time on their voyage across the sea.

How does this system work? According to the 20th-century British explorer and navigator Sir Thomas Gladwin, for each particular voyage the sailor selects an etak reference island. Usually it lies 50 miles (80 km) to one side of the direction of destination. Then star compass bearings are singled out from memory. In between a navigator recognizes other reference stars that the island will pass beneath as it glides backward with respect to the canoe. These mark out etak segments of the voyage. Like charting out a flight from New York to Los Angeles into segments based on successive states you pass over, the navigator divides up the time-space of the whole trip into etak segments so that he can estimate what portion of the trip has been completed.

The blending of time and space inherent in ocean or, for that matter, mountain travel is further reflected in another exotic device once used in the Marshall Islands—a kind of map that involves both space and time. First described by Western navigators in the 19th century, a Melanesian stick chart is a complex communication scheme that is as different from Western charts and maps as anyone can imagine. Numerous examples exist, and thanks to some good early ethnological accounts, we know a great deal about how they work.

By using a stick chart, Marshallese navigators transformed their sensory impression of the watery environment into a concrete mnemonic scheme that helped them deal with the central problems of navigation. Unlike the Caroline Islanders' star compass, which was committed to memory, the stick chart was a real technological artifact. It consisted of a network of straight and curved reeds bound together to form a flat frame about 30 inches by 30 inches (75 cm by 30 cm) that could conveniently be held in the hand.

In one representation the seafarer thinks of the stick chart as a map to tell where to proceed. In this interpretation, the straight and curved lines represent the course, while their interstices signify the location of the islands. These are often symbolized by cowrie shells fastened to the reed matrix. But another interpretation is that the navigator can use the curves and the intersections on the chart to call to memory the behavior of wave patterns experienced on the ocean surface. Refraction and reflection of waves by neighboring islands, many of them well out of view, produce certain aberrant patterns upon the surface of the sea. These disturbances cause the boat to pitch and yaw ever so slightly and in such a manner that, while the patterns cannot be seen with the eye, mild movement can be felt by the body of the sailor. These same interference patterns caused by oceanic swells have only recently been detected by modern oceanographers by means of satellite photography.

For the Pacific navigator, ocean wave interference patterns, like the stars, served as directional aids. One example is the intersection of eastward- and westward-moving swells refracted by an individual island. Lines that trace intersecting successive swells produce a disturbance that increases in intensity as one approaches the island that produces it. Thus, should one happen to be voyaging to or from the island that produces the disturbance, the line of nodes becomes the navigator's course. In the example shown on page 155, the vertical axis of the stick chart represents such a course, and the bent reeds symmetrically arranged about the axis signify the eastward- and westward-moving swells. In this case the island would lie at the center of the chart. It is important to realize that the two interpretations of the stick chart, both as a map and as a taxonomy of disturbance patterns, were often engaged simultaneously. Like *etak*, stick-chart navigation is a very difficult notion to fathom for those of us who are used to separating time and distance.

The earth and sky environment always seems to provide the anchor, the shaping for the direction of human thought. Like linear constellations, the stick chart was invented in response to the phenomena of nature, in this case the active behavior of ocean swells unique to the Marshall Islands. Here a navigational corridor is produced by two chains of islands that run 30 to 60 miles (50 to 100 km) apart for nearly 600 miles (950 km). Within this region many different kinds of interference patterns occur. Because navigation is the most valued activity among the people who populate these tiny, Pacific atolls, navigators needed to develop a means of recording and communicating practical knowledge about the ocean environment. Not surprisingly, the materials involved in human expression consisted of reeds and shells, the most readily available products on the island shores.

One of the great mysteries of the many cultures that comprise Oceania concerns their origin. Where did settlers come from in the first place? How and when did they arrive from the mainland? Deep-sea voyaging across the Southeast Asian Archipelago seems very difficult. Modern anthropologists and adventurers like David Lewis and Ben Finney—acting in the spirit of the legendary Thor Heyerdahl—have used native technological skills to reenact such voyages themselves, clearly demonstrating that they could have been made. But why travel in the first place? There would have been no shortage of practical motives beyond pure adventure for conducting long-distance trips across vast expanses of water: to trade material goods, to obtain food and other items from uninhabited islands, to conduct raids or warfare and to extend a chiefdom's sovereignty. One group of Cook Islanders, for example, was said to have traveled more than 200 miles just to procure supplies of birds' eggs. Whenever economic disruption disturbed the mainland, it would prompt a migration to the more sparsely inhabited, or totally uninhabited, islands even farther off the coast. It is well known that the city-state of Venice was founded as the result of just such a process, after Lombard invaders began to intrude upon the resource base of the northern Latin cultures which, for centuries, held sway over the rich fishing grounds in the Venetian lagoon. Indeed, the development of celestial navigation as Oceania's highest-prized human skill may in fact have been a cultural adaptation by mainland people essential for their future survival in a watery environment.

Can the study of the astronomy of cultures nearer the mainland offer a clue to whether people diffused down the Southeast Asian Archipelago carrying ancient sky traditions to be altered and adapted by later generations with a different set of needs and conditions? The practice that we encounter in the Pacific islands of pivoting the astronomical reference frame about the overhead point also exists in Java (latitude 7°S). Early Dutch explorers had learned of a gnomon or shadow-casting device that was used to partition the year in a unique way. The cruciform plate of the device is fashioned to reckon the start of the year from the June solstice, the first day of winter in the Southern Hemisphere. The so-called rustic year is segmented into 12 months of unequal duration, ranging from 43 days (the first, sixth, seventh, and twelfth months) to 23 days for certain months in between. Now, these unequal time intervals are precisely the ones obtainable if you use as the guiding principle equal units of length traveled by the shadow cast by the tip of the gnomon at noon. The passage of the sun through the zenith would be the pivotal point for dividing up the year.

In the latitude of central Java (Indonesia), a very special condition occurs in gnomonic geometry to make all of this possible. When the sun stands on the meridian at noon north of the zenith on the June solstice, the shadow length, measured to the south of the base of the vertical pole, is exactly double the length that you find when the sun at noon lies at the December solstice south of the zenith, then projecting the shadow to the north. Javanese chronologists halved the shorter segment and quartered the longer, and this is how they produced their bizarre 12-month calendar. As in most cultures, the number 12 probably was chosen as a way to try to fit a whole number of full moons into the yearly scheme. But because the sun moves more slowly at the solstices, we should expect the first and last months, along with the sixth and seventh months, when the shadow approaches its southerly and northerly maximal extensions respectively, to be those of greatest duration. On the other hand, months defined by equal shadow lengths when the sun is near the equinox ought to be relatively brief. This is exactly what happens in the Javanese time scheme. The result is a calendar that consists of 12 months of variable length, ranging from 23 to 41 days. Once they had captured the celestial symmetry, calendar-keepers later altered particular intervals by one or two days either way in order to force them to correspond more precisely to agricultural practices and other human activities.

The development of the ancient Indonesian calendar is traceable even further back in history and the record shows just how adaptable to human needs the sky and its denizens can be. The existence of these strange months of non-uniform length was first reported in the literature more than a century ago by a British traveler named John Crawfurd, who was totally baffled at their occurrence. Though he refers their origin to the *banchet* or gnomon, he never explains how the intervals were determined. That task was left to the later Dutch anthropologists. Crawfurd only comments that the first 10 months carried the names of the ordinal numerals in the local language and he suggests that two more months were probably added later by the Brahmins who entered Java from India and sought to make the local calendar correspond to their own. Evidently, the Brahmins threw the beginning of the year from the first back to the 11th month, which corresponds roughly with our month of April, in order to make local time fit the Hindu year. Still later, the solar zenith pivot and a gnomonic device to carve up time may have served the purpose of tying regional calendars together.

How else were the timekeeping systems of various villages actually linked to a standard time? There once was confusion about when to begin sowing, and

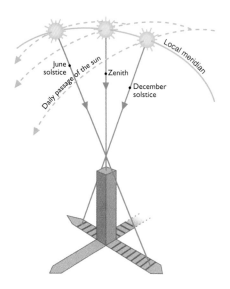

The Javanese used an unusual sundial that divided up the year by following the noontime shadow of the gnomon (vertical stick) over equal intervals of length on the base plate.

Crawfurd relates that "anciently it was regulated by the stars, and particularly by the appearance (heliacal rising) of the *Bintang, Buriak*, or Pleiades." When the lunar year of the Hegira was introduced, the season gradually slipped out of joint with the timing mechanism because of the 11-day lapse between the lunar and solar years. One ingenious method devised by rice farmers of the Kenhay-Kayan complex for determining when to plant by the stars consisted of pouring water into the end of a bamboo pole. Then they pointed it at a particular star, thus allowing some of the water to spill out. When the stick was held vertically, if the water level corresponded to a predetermined mark, then it was time to plant.

There is a parallel between the Javanese calendar and the one developed by the Trobriand Islanders who live to the east, near Papua New Guinea. Half a century ago, the British anthropologist Sir Edmund Leach recognized that New Year's Day in the Trobriand Islands is marked not by a solar event, nor even an astronomical one, but rather by the appearance of a worm. But the palolo worm, an inhabitant of coral reefs, is linked to the sky by its "circa-lunar" rhythm. Once a year, for three or four nights during spawning, the posterior parts of the worm wriggle dramatically on the surface of the water and expel their reproductive fluids.

Every year, following the appearance of the full moon between October 15 and November 15 (our time), the spawning marine annelid is seen on the surface of the sea at the southern extremity of the island chain. Trobrianders name this important month *Milamala* after the worm, and they hold a great festival in its honor to inaugurate the planting season, a celebration during which they dine on this edible delicacy. Anthropologists have discovered that this same feast was held one month earlier in the main part of the island just north of the area, two months earlier in the outlying islands to the south, and three months earlier to the east.

Trobriand space and time join in Milamala, for this period is considered on the whole as a set of four months that are broken down regionally among different island groups. How does the calendar work in practice? Consider the month-naming sequence in four districts, A, B, C, and D. Suppose everybody agrees that district D will be the control or checkpoint, because the people in district D are the fishermen who actually observe the worm. Now, once the worm appears, the people of district D call the full moon "just past Milamala"—that is, they name it retroactively. But what if the worm is late, as we can expect it to be in a lunar calendar with a fixed number of months? In that case, the people of the area D simply name and celebrate an extra Milamala month on the spot. This would be a bit like those of us in northern climes assigning another month of December if snow did not arrive in time for Christmas. The whole process is distinctly reminiscent of the way the Mursi of East Africa link the moon to other aspects of nature, like the cresting of their river. However, such action puts all the other districts out of line relative to D. But this does not upset the Trobrianders any more than the cresting of the

Omo River in different places and at different times bothers the Borana. They respond later (sometimes even up to a year or two later!) by doubling their own Milamala season. In due course, each district decides what it will do and when it will do it based upon their own separate time-checking schemes. Just as some states in the United States can choose not to adopt daylight-saving time, local autonomy reigns. And the overall discourse about time keeps people talking to one another.

One curiosity about the Trobriand system clearly visible in the system is that the whole of the territory completes a 12- or a 13-month lunar cycle, while no given area actually counts more than 10 months. Once again we encounter that familiar "time that lies outside of the regular calendar." In this case, the "time-out" is a free-floating adjustable period tied to the occurrence in nature of a singular event that resets the calendar—the appearance of the palolo worm.

Calendars with short years occur in other astronomies. Remember the Inca system that began and ended the year with the appearance and disappearance of the Pleiades, or the ancient form of our own, the Roman calendar which once began with March, the month of the equinox, and ended with the eighth, ninth, and tenth lunations, as their Latin names betray—October, November, December. As for the rest of the year, following the tenth month there was a gap of two, maybe three months until the next cycle, timed by some now unknown event that signaled the equinox. Christian Easter is a modern survival of the start of the year. This uncounted interval corresponded to a temporal limbo when fields lay fallow, that time-out borderland in which the farmers waited patiently for nature to signal the awakening of spring, to rekindle the cycle of life. Although we have come to think of the 365-day solar year as somehow categorically correct, these old moon-based calendars, especially the four-district system of the Trobrianders, illustrate how a minimum of systematic knowledge about the passage of natural events can be organized into a complex and workable calendrical system.

Modifying nature's time spans to achieve greater social unity is an activity common to all cultures, and the Julian and Gregorian reforms are not the only sophisticated attempts in the civilized world devised by clever astronomers who create their own kind of time to regulate human behavior. While many of the details of Trobriand or Javanese calendar reform have not survived or still lie hidden, we can well imagine that their chronological systems underwent change both as a result of attempts to formalize over a wider segment of the population the activities associated with fishing or rice planting and those tied to the sun, moon, and stars. Indigenous contact with India and later with the West resulted in further calendar reforms. There is little doubt that such changes were motivated mainly by attempts in Indonesian societies to expand the base of economic resources. At least they can scarcely be attributed to purely ideological struggles about the nature of time.

This 17th-century star map from a celestial atlas by Andreas Cellarius is considered one of the most beautiful of all celestial charts. It was the process of being born.

made at a time when modern astronomy was in

11

A JEWEL WITH MANY FACETS

We have explored the archaic depths of the most ancient of all sciences. From the earliest material records in the Paleolithic caves of France—where carved bones recorded the first attempts of our Neolithic ancestors to devise a calendar—to still-told tales around a Navajo campfire in the Four Corners area of the United States, we have sampled a variety of ways through which people have discovered and expressed their relationship to the sky. The modern dictionary definition of astronomy should now seem obsolete and one-dimensional. It depends too much on high technology for acquiring astronomical information, and too much on writing and tabulation as a way of recording celestial imagery.

The astronomy we have explored goes well beyond Ptolemy's *Almagest*, Islam's star tables, Babylon's cuneiform clay tablets, and the Mayan codices. The ancient astronomers' records we have encountered display a veritable rainbow of colorful categories. Hesiod's oral poem dramatized the early Hellenic farmer's deep awareness of the connection between stars and winds, sun and birds, moon and the sowing of seed. Embedded within the very structure and orientation of Maya, Inca, and Chinese cities lie great ideas of royal, astronomically minded architects—beliefs about the creation of the universe and how the mountain-sky periphery surrounding the ceremonial center bound society together. Pawnee and Yekuana communal residences, and places of assembly like England's Stonehenge and Ohio's Octagon Mounds, became time charts for scheduling the rituals that needed to be performed regularly and repeatedly within nature's theater in order to renegotiate the bonds that united people with their gods. Mimbres ceramics and Maya carved stelae tell mythological stories of lunar rabbits and real-life blood oaths between kings, queens, and their guiding sky spirits. If we separate these unwritten forms of human expression from the books and tables of the later Western tradition, then our understanding of the underpinnings of ancient science will remain forever incomplete.

Astronomy was *lived* as intently as it was practiced, and much of what our predecessors viewed was devoted to purposes that, today, we would regard as religious rather than scientific. When I chip away to the bedrock of my chosen discipline, I always seem to find astrology lying at its substrate. Is it because deep down we all believe that we live in an animate world and always wish to remain in constant dialogue with it? For most of human history we believed in celestial deities, many of whom behaved like petulant, if super-powerful, versions of ourselves. If we disregard the metaphysical side of our ancestors' outlook, and focus only on those aspects of their astronomy that closely resemble our own—at the same time discarding astrology and mythology on the trashheap of mysticism—then I think we may be missing an important part of humanity's outlook on the universe.

The world's first astronomers seem very remote from their modern counterparts, who practice that dictionary definition of astronomy. Today's skywatching is an activity we conduct largely by remote sensing. We say we "experience" the atmosphere of Jupiter and the volcanoes on the molten surface of that planet's satellite, Io, even though we have been to neither place in the flesh. With the aid of enhanced television imagery derived from remotely beamed and processed electronic signals, we tell of the things we "see" beneath Venus's cloudtops or deep within the solar inferno. But when the ancient astronomers saw Venus resurrected or the sun progress to its standstill, they experienced the universe directly, without technological enhancement. Little came between eye and sky in those days; only simple devices like Greek and Roman sundials, Mexican crossed sticks, Teotihuacán pecked circles, or Egyptian merkhet.

The more we probe, the more we discover that the ancient astronomy that came out of Stonehenge, Uruk, or Tenochtitlán was not so different from the astronomies of contemporary non-Western cultures like the Yekuana and Desana of South America, the Borana and Mursi of Central Africa, or the Trobrianders and Puluwatese of Oceania. Past or present, only the West seems to have taken a different route, one we have traveled for but a few centuries. For us, the universe is inanimate and separated from the conduct of human affairs. For them, it is alive and intimately attached to every activity they undertake.

Would we regard Mesoamerican calendar-keeping or Melanesian celestial navigation as examples of scientific endeavor? True, neither of these cultures used a sophisticated technology based on models and machinery such as armillary sphere or astrolabe—the tools of those earlier astronomers of whom we are the direct descendants. Neither was there a complex geometry or algebra that served as the language for expressing what Netzahualpilli—an astronomer from Texcoco, the city that rivaled Tenochtitlán—witnessed every night from the observatory he had had constructed on the roof of his palace. Did the priests who wrote the Mayan codices care about scientific theorizing based on abstract, non-human concepts such as force, mass, and acceleration? Would they be indifferent to our Big Bang theory of the origin of the universe? We could argue that the 3-D navigator's star compass used in Oceania before any contact with the West is far more abstract than the magnetic compass common to China and the West, and that the Mayan or Tuamotuan models of a layered heaven are as theoretically based as an astrophysicist's model of the interior of a planet or the initial expansion of the universe. If we want to think of those early efforts as truly scientific, we would also need to ask whether the ancient astronomers believed—as we do—in the idea of progress as a way to ultimate truth. And we would need to inquire whether they ever sought to test and improve their astronomical models in the face of newer observations, as was done in the West.

But is it reasonable even to think that the people of ancient China, India, the Andes, the Amazon, and Mexico, all with their own diverse customs and histories, considered nature as we do? Unlike our own, their pasts did not necessarily consist of feudal state, imperial expansion, an artistic Renaissance, a religious reformation and counter-reformation, eventually followed by the sudden and rapid flourishing of technology and democratic principles of government. And even if they did, could we dare anticipate that the outcome of the dialogue between nature and culture after centuries of mixing and blending of people and ideas—much of it in long-term isolation from the West— would produce the exact, quantitative astronomical science we know today?

In this modern view of a full moon we see the same shape that skywatchers from many different cultures across many hundreds of years have identified as a rabbit. Did such ideas spring from one culture and spread throughout the world, or did they occur independently to all those who looked to the heavens for an understanding of the world around them?

While astronomy's uses throughout the ancient world, from the practical to the esoteric, seem very diverse, some common denominators do surface when we cut across the globe. Why was every culture that we encountered so obsessed with fabricating commensurable time loops? The eight-year cycle named *octaeteris* by the Greeks also was picked up by the Maya. In China and India, in Islam and the Middle East, astronomers in charge of the calendar were constantly preoccupied with finding the magic numbers through which the gears of planetary time units could be meshed together. Why did the Desana, the Maya, and the Greeks choose to invent a zodiac and why did all three sky bands consist basically of a parade of animals? Why did Barasana and Pawnee, Gilbertese and Yekuana all pattern their domiciles after the great cosmic house that protected their society and defined their home place? And how could civilizations, developed in virtual isolation from one another, all see a rabbit in the moon?

Can a single culture's ideas have diffused round the world long ago? While there is little question that people migrated down and across the Southeast Asian Archipelago, the overwhelming weight of archaeological and other scientific evidence argues against any large-scale diffusion of people and ideas from North to South or East to West and vice versa. And yet, underlying all the world's ancient astronomies, there remain those tantalizing common denominators that can be traced in Old and New World cultures. Perhaps some basic ideas about the sky were transmitted across the Bering land bridge via early migratory patterns of archaic peoples. For example, a fixed polar-based astronomy for finding the way or the tradition of a symbolic four-dimensional universe are deeply embedded archetypal concepts that can endure in the mind for a long time, passing from generation to generation by word of mouth.

Personally, I feel more comfortable believing that most astronomical problems, and many of their common solutions, would occur independently to anyone who carefully followed events in the heavens. Regardless of language and culture, determining how to make cycles of full moons fit with the annual course of the seasons marked by the sun is part of a human discourse with the sky—a discourse concerned both with predicting the future and with creating order where the eye first sees chaos.

We need to avoid sweeping whatever may seem foreign and irrational about the sky under the rug of ancient astronomy, and we must resist the inclination to label the exotic and unfamiliar as worthless because it hindered or retarded the development of astronomical exact science as we know it today. Instead, we should try to comprehend what happens above as witnessed through the eyes—and contemplated through the minds—of the diverse societies and cultures who inhabit the space here below.

We are further now from forming a simple definition of astronomy than when we first set out. Maybe we have learned more about what astronomy is not, than about what it is. Mursi astronomy, with its built-in disagreements, proved to us that you do not have to know how to write to be an astronomer and that a calendar need not be a piece of paper. The East Indian way of viewing the sky has taught us that cosmology need not consist of a model that explains physical phenomena. And our Barasana observations demonstrated that astronomy is not just a cognitive system and a mere body of knowledge; it is also an organized way of knowing that gives people a formula for action, a means of exerting human effort to keep the cosmos in orderly motion, a scheme to telescope their time and their place to cosmic proportions.

In this sense, whether societies be organized hierarchically or not, the goal is the same. The Barasana, or for that matter the Inca and the Aztecs, really do share our modern scientific attitude, which strives to reach out to the stars—not only because they are there, but also to master the predictive laws that hold the power to tell where they will be in the future. Whether we use prayer to commune with the sky or the natural laws of mathematics, what lies at the very foundation of our motivation to study astronomy is the will to understand and control the cosmos—with or without the gods, limitless or limited.

So, the question that we have put to the ancient astronomers—why look at the sky?—has returned a kaleidoscope of responses, as many and as illuminating as the facets on a finely cut diamond. The stars that shine on every point of the diamond in the sky are there to orient us as individuals, and to help order our society as a whole. They are also there to give meaning to the human spirit. What better place to find the clues and patterns that offer answers about the meaning of life than in the ever-dependable, supportive, and nurturing world around us?

REFERENCES

CHAPTER 1

For those who want to get acquainted with the nighttime sky, monthly sky maps and lists of naked-eye celestial phenomena can be found in *Sky & Telescope* magazine, published by Sky Publishing Corp. The Western common sense concepts in astronomy are clearly explained in the classic *The Fabric of the Heavens* by S. Toulmin and J. Goodfield. I drew significantly on this text for my first three chapters.

AVENI, A.F. 1980 *Skywatchers of Ancient Mexico* University of Texas Press, Austin.

AVENI, A.F. 1981 Tropical Archaeoastronomy, *Science* 213: 161-171.

AVENI, A.F. 1989 *Empires of Time*, Basic Books, Inc., New York.

KRUPP, E.C. (EDITOR) 1978 *In Search of Ancient Astronomies* Doubleday & Company, Inc., Garden City. pp 1-37

CHAPTER 2

I drew much of the resource material on Stonehenge and the megalithic sites of Great Britain and Northwestern Europe from articles published in the British *Journal for the History of Astronomy* and its occasional supplement, *Archaeoastronomy*. At a more popular level I recommend C. Chippindale's *Stonehenge Complete*. R. Castleden's book is an excellent account of what the societies of the builders of these sites were like, based upon the archaeological record. Aubrey Burl's *Stone Circles of the British Isles* remains the best compilation of the full archaeological documentation. My own *Empires of Time* gives a detailed discussion of celestial timekeeping before the advent of writing as we know it.

AVENI, A., H. HARTUNG AND J.C. KELLEY 1982 Alta Vista (Chalchihuites), Astronomical Implications of a Mesoamerican Ceremonial Outpost at the Tropic of Cancer, *American Antiquity* 47, 316-35.

BURL, A. 1976 *Stone Circles of the British Isles,* Yale University Press, New Haven.

CASTLEDEN, R. 1987 *The Stonehenge People*, Routledge & Keagan Paul, London.

CHIPPINDALE, C. 1983 *Stonehenge Complete*, Cornell University Press, Ithica.

FRAZER, R.M. (EDITOR) 1983 *The Poems of Hesiod*, University of Oklahoma Press, Norman.

HAWKINS, G. 1965 *Stonehenge Decoded*, Doubleday & Company Inc., Garden City.

MARSHACK, A. 1972 *The Roots of Civilization*, McGraw-Hill, New York.

MURRAY, W.B. 1982 Calendrical Petroglyphs of Northern Mexico in *Archaeoastronomy in the New World*, ed. A.F. Aveni, Cambridge University Press, Cambridge.

CHAPTER 3

There are many excellent texts that deal with Greek astronomy. I can recommend D.R. Dicks' *Early Greek Astronomy to Aristotle,* which I employed in this chapter along with A. Pannekoek's, *History of Astronomy*. I also drew some of my material from a series of essays by D. J. de Solla Price entitled *Science Since Babylon* (1961 Yale University Press, New Haven). D. Schmandt-Besserat's *Before Writing* (2 Vols) (1992, University of Texas Press, Austin) gives an excellent account of how the Babylonian cuneiform system, which contained some of the earliest astronomical records from the Fertile Crescent, developed. J. P. Vernant offers a lucid explanation of the social origins of Greek geometrical theorizing in *Myth and Thought Among the Greeks* (1983, Routledge & Keagan Paul, London). See also selected essays in Brecher and Feirtag's *Astronomies of the Ancients* and E.C. Krupp's *In Search of Ancient Astronomies*.

AABOE, A. 1974 Scientific Astronomy in Antiquity in *The Place of Astronomy in the Ancient World*, ed. F.R. Hodson, Oxford University Press, London, 21-42.

BRECHER, K. & M. FEIRTAG 1979 *Astronomies of the Ancients*, MIT Press, Cambridge.

BOWEN A. & B. GOLDSTEIN 1989 Meton of Athens in *A Scientific Humanist*, ed. E. Leichty et al, University Museum, Philadelphia.

DICKS, D.R. 1970 *Early Greek Astronomy to Aristotle*, Thames and Hudson, Bristol.

GIBBS, S. 1979 The First Scientific Instruments in *Astronomy of the Ancients*, ed. K. Brecher & M. Feirtag, MIT Press, Cambridge. pp 39-59.

KRUPP, E.C. 1977 Astronomers, Pyramids and Priests in *In Search of Ancient Astronomies*, ed. E.C. Krupp, Doubleday & Company, Inc., Garden City. pp 203-239.

NEUGEBAUER, O. 1983 *Astronomy and History Selected Essays* Springer-Verlag, New York. 196-209, 211-212 (On how to orient a pyramid).

O NEIL, W.M. 1986 *Early Astronomy from Babylonia to Copernicus.* Sydney University Press, Sydney.

PANNEKOEK, A. 1961, *A History of Astronomy.* Oxford University Press, London.

PARKER, R.A. 1974 Ancient Egyptian Astronomy *The Place of Astronomy in the Ancient World*, ed. F.R. Hodson, Oxford University Press, London, pp 51-65.

PRICE, D.J. DE SOLLA 1974 *Gears from the Greeks.* The American Philosophical Society, Philadelphia.

SACHS, A. 1974 Babylonian Observational Astronomy. *The Place of Astronomy in the Ancient World*, ed. F.R. Hodson, Oxford University Press, London. pp 43-50.

SCHMANDT-BESSERAT, D. 1986 An Ancient Token System, The Precursor to Numerals and Writing. *Archaeology* 39 (6), 32-39.

TOULMIN, S. & J. GOODFIELD 1961 *The Fabric of the Heavens.* Hutchinson & Co., London. .

CHAPTER 4

I discuss the relationship between astrology and astronomy in *Conversing with the Planets* (1992 Times Books, New York). Seyyed Hussein Nasr's beautifully illustrated book was the source of much material that went into the production of this chapter (see also his *Introduction to Islamic Cosmological Doctrines*, 1978 Thames and Hudson Ltd., London). I also used some of the more technical publications in *From Deferent to Equant: A Volume of Studies in the History of Science in the Ancient and Medieval Near East in Honor of E.S. Kennedy* (Annals NY Academy of Sciences Vol. 500, 1987). The development of Islamic astronomic instruments is covered in Singer's *History of Technology*. An especially clear account of how an astrolabe works is offered in J.D. North's *Scientific American* article, and in O'Neil, pp. 117-132.. Allen's *Star Names: Their Lore and Meaning* is a fairly reliable account of the Arabic origin of stellar nomenclature. The best discussion of the q'ibla system of celestial alignment is given in the many publications of D. King; see especially his essay in *Ethnoastronomy and Archaeoastronomy in the American Tropics*, Annals NY Academy of Sciences Vol 385.

ALLEN R. 1963 *Star Names, Their Lore and Meaning.* Allen & Unwin, New York..

GINGERICH, O. 1987 Zoomorphic Astrolabes and the Introduction of Arabic Star Names into Europe, Annals NY Academy of Sciences Vol. 500: 89-104.

KING, D. 1982 Astronomical Alignments in Medieval Islamic Religious Architecture, Annals NY Academy of Sciences, Vol. 385, 303-312.

NASR, SEYYED HUSSEIN 1976 *Islamic Science: An Illustrated Study.* World of Islam Festival Publishing Co., Westerham, U.K..

NORTH, J.D. 1974 The Astrolabe. *Scientific American* 230, 96-106.

WAERDEN, B.L. VAN DER 1987 The Heliocentric System in Greek, Persian and Hindu Astronomy, Annals NY Academy of Sciences Vol 500: 525-545.

CHAPTER 5

One of the clearest accounts of the ancient Chinese philosophy of astronomy that I can recommend is E. Schafer's *Pacing the Void.* I also drew on J. Needham's celebrated multi-volume work *Science and Civilization in Ancient China.* The place of astronomy and cosmology in the making of the ancient Chinese city is beautifully articulated in P. Wheatley's *Pivot of the Four Quarters.* At a more popular level see the illustrated essays of E.C. Krupp in the 1980s *Griffith Observer.* A special thanks to Dr. Krupp, who commented on an earlier draft of this chapter.

HO PENG YOKE 1966 *The Astronomical Chapters of the Chin Shu.* University of Malaya Press, Kuala Lumpur.

KRUPP, E.C. The Cosmic Temples of Old Beijing. *World Archaeoastronomy*, ed. A.F. Aveni, Cambridge University Press, Cambridge. 65-74.

MORGAN, T. 1980 The Burmese Era and Ancient Astronomy in Southeast Asia. *Archaeoastronomy* The Bulletin of the Center for Archaeoastronomy III (2): 20-21.

NEEDHAM, J. 1954-1988 *Science and Civilisation in China.* Cambridge University Press, Cambridge.

NEEDHAM, J. 1974 Astronomy in Ancient & Medieval China. *The Place of Astronomy in the Ancient World* ed. F. Hodson, Oxford University Press, London, pp 67-82.

NIVISON, D. 1977 The Origin of the Chinese Lunar Lodge System. *World Archaeoastronomy*, ed. A.F. Aveni, Cambridge University Press, Cambridge, pp 203-218.

PANKENIER, D. 1982 Early Chinese Positional Astronomy. *Archaeoastronomy* The Bulletin for the Center for Archaeoastronomy V (3): 10-19.

SCHAEFER, B. 1983 Chinese "Astronomical" Jade Disks: The Pi. *Archaeoastronomy* the Bulletin of the Center for Archaeoastronomy VI, 99.

SCHAFER, E. 1977 *Pacing the Void.* University of California Press, Berkeley.

SIVIN, N. in Chinese Archaeoastronomy: Between Two Worlds. *World archaeoastronomy*, A.F. Aveni, Cambridge University Press, Cambridge, 55-64.

XI ZEZONG 1984 New Archaeoastronomical Discoveries in China. *Archaeoastronomy* The Bulletin for the Center for Archaeoastronomy VII, pp 34-45.

XU ZHENTAO, K., K.K.C. YAU & F.R. STEPHENSON The Shang Dynasty Oracle Bones. *Archaeoastronomy* Supplement to Journal for the History of Astronomy 14, S 61.

WHEATLEY, P. 1971 *Pivot of the Four Quarters.* Aldine Publishing Company, Chicago.

CHAPTER 6

Material on indigenous African astronomy and cosmology is relatively sparse. The subject matter I employed in this chapter can be found in selected essays published in the *Archaeoastronomy Bulletin*. S. Blier's treatment of astronomical and cosmological symbolism in Batammaliba architecture is highly recommended, while Ruggles and Turton & Ruggles give the best treatment of the role of astronomy in calendar keeping. I discuss some of their work in addition to that of E.E. Evans Pritchard at a more basic level in my *Empires of Time*. My hypothetical model of how to predict the weather by the tilt of the crescent moon was inspired by Deirdre LaPin's short film on the Ngas entitled *Sons of the Moon*.

BLIER, S. 1982 African Cosmology. *Archaeoastronomy* The Bulletin for the Center of Archaeoastronomy V(3): pp.40-41.

BLIER, S. 1982 African Cosmology, Astronomy and World View. *Archaeoastronomy* The Bulletin of the Center of Archaeoastronomy V(4): pp.4-5.

BLIER, S. 1987 *The Anatomy of Architecture.* Cambridge University Press, Cambridge.

LAPIN, D. 1982 (unpub.) Moon, Science & Symbolism on the Bauchi Plateau.

ROBERTS, A. 1981 Passage Stellified: Speculation upon Archaeoastronomy in Southeastern Zaire. *Archaeoastronomy* The Bulletin of the Center for Archaeoastronomy IV (4): pp.26-37.

RUGGLES, C.L.N. 1987 The Borana Calendar: Some Observations. *Archaeoastronomy* Supplement to Journal for the History of Astronomy 11: 35.

TURTON, D. & C.L.N. RUGGLES 1978 Agreeing to Disagree: The Measurement of Duration in a Southwestern Ethiopian Community. *Current Anthropology* 19,(3) 585-600.

CHAPTER 7

By contrast to the previous chapter, the fount of published material on Mesoamerican astronomy and calendars is rather large. Most of the basic references prior to 1980 are given in my *Skywatchers of Ancient Mexico.* The most recent detailed account of our knowledge of the Maya calendar is, in my opinion, John Justeson's lengthy essay in my edited *World Archaeoastronomy.* A popular account of how astronomy fit into the lives of the Maya royalty can be found in L. Schele and D. Freidel's two books. Concerning the people of highland Mexico, especially the Aztecs of Tenochtitlan, see D. Carrasco's edited *To Change Place* (University Press of Colorado, Niwot) and *The Great Temple of Tenochtitlan* essays by J. Broda, D. Carrasco, and E. Matos Moctezuma (1987 University of California Press, Berkeley). J. B. Carlson's beautifully illustrated article in *National Geographic Magazine* (March, 1990 pp. 76-107) is recommended reading for this chapter as well as the next two.

AVENI, A. 1977 Astronomy in Ancient Mesoamerica. *In Search of Ancient Astronomies*, ed. E.C. Krupp, Doubleday & Company, Inc., Garden City, pp. 165-202.

AVENI, A.F. 1980 *Skywatchers of Ancient Mexico.* University of Texas Press, Austin.

AVENI, A.F. 1986 Non-Western Notational Frameworks and the Role of Anthropology in Our Understanding of Literacy. *Toward a New Understanding of Literacy*, ed. M. Wrolstad and D. Fisher, Praeger, New York, pp. 252-284.

FREIDEL D. & L. SCHELE 1993 *Maya Cosmos.* William Morrow, New York.

GOSSEN, G. 1974 A Chamula Calendar Board from Chiapas, Mexico. *Mesoamerican Archaeology: New Approaches,* ed. N. Hammond, Gerald Duckworth and Co. Ltd., London, pp. 217-253.

JUSTESON, J. 1989 Ancient Maya ethnoastronomy: an overview of Hieroglyphic Sources. *World archaeoastronomy*, ed. A.F. Aveni, Cambridge University Press, Cambridge, pp. 76-129.

MATOS MOCTEZUMA, E. 1988 *The Great Temple of the Aztecs.* Thames and Hudson Ltd., London.

MILLER, M. 1986 *The Murals of Bonampak.* Princeton University Press, Princeton.

SCHELE, L. 1977 Palenque: House of the Dying Sun. *Native American Astronomy,* ed. A.F. Aveni, University of Texas Press, Austin, pp. 42-56.

SCHELE, L. & D. FREIDEL 1990 *A Forest of Kings.* William Morrow, New York.

SCHELE, L. & M. MILLER 1986 *Blood of Kings, Dynasty and Ritual in Maya Art.* George Braziller, Inc., New York.

WICKE, C. 1984 The Mesoamerican Rabbit in the Moon: An Influence from Han, China? *Archaeoastronomy* The Bulletin of the Center for Archaeoastronomy VII, 46-55.

CHAPTER 8

J. Carlson and J. Judge's edited Maxwell Museum text was a source of a lot of the material I compiled on the archeoastronomy of the U.S. Southwest, as was R. Williamson's book *Living the Sky.* Astronomical myths and legends across North America are discussed in Williamson and Farrer's *Earth & Sky.* In the preparation of this chapter, I also drew on selected essays in the Brecher and Feirtag as well as the Krupp volume and from my own edited *Native American Astronomy and Archaeoastronomy in Pre-Columbian America,* both published by the University of Texas Press (see also the many published works in *Archaeoastronomy*, Supplement to Journal of the History of Astronomy).

CARLSON, J. & J. JUDGE (ED.) 1987 *Astronomy & Ceremony in the Prehistoric Southwest.* Maxwell Museum, Albuquerque.

CHAMBERLAIN, VON DEL 1982 *When Stars Came Down to Earth.* Ballena Press/Center for Archaeoastronomy Cooperative Publication, Los Altos.

EDDY, J. 1977 Archaeoastronomy of North America *In Search of Ancient Astronomies*, ed. E.C. Krupp, Doubleday & Company, Inc., Garden City, pp. 133-163.

HIVELY, R. & R. HORN 1982 Geometry and Astronomy in Prehistoric Ohio. *Archaeoastronomy* Supplement to Journal for the History of Astronomy, 4. pp. S1-S20.

HIVELY, R. & R. HORN 1984 Hopewellian Geometry and Astronomy at High Bank. *Archaeoastronomy,* Supplement to Journal for the History of Astronomy. 7, 85; 4 1.

KRUPP, E.C. 1977 Cahokia, Corn, Commerce, and the Cosmos. *Griffith Observer* 41 (5) pp. 10-20.

MARSHACK, A. 1989 North American Calendar Sticks. *World Archaeoastronomy,* ed. A.F. Aveni, Cambridge University Press, Cambridge, 308-324.

MCCLUSKEY, S. 1989 Lunar Astronomies of the Western Pueblos. *A World Archaeoastronomy,* ed. A.F. Aveni, Cambridge University Press, Cambridge, pp. 355-364.

PARSONS, E.C. (ED.) 1936 Hopi Journal of Alexander M. Stephen. *Contributions to Anthropology* Vol. 23 Columbia University, New York.

ROBBINS, R. & R. WESTMORELAND 1991 Astronomical Imagery & Numbers in Mimbres Pottery. *The Astronomical Quarterly* 8, 65-88.

SOFAER, A., V. ZINSER & R.M. SINCLAIR 1979 A Unique Solar Marking Construct. *Science* 206: 283-291.

WILLIAMSON, R. 1984 *Living the Sky.* Houghton Mifflin, Boston.

WILLIAMSON, R. & C. FARRER (ED.) 1992 *Earth & Sky.* University of New Mexico Press, Albuquerque.

ZEILIK, M. 1989 Keeping the Sacred and Planting Calendar: archaeoastronomy in the Pueblo Southwest. *World Archaeoastronomy,* ed. A.F. Aveni, Cambridge University Press, Cambridge, pp. 143-166.

CHAPTER 9

Much of the material on Andean and other indigenous South American astronomies is difficult to access because it is integrated into anthropological studies of the living cultures and much of it is in Spanish. Readers might consult, for example, books by S. Fabian on *Bororo Astronomy and Calendar* (University of Florida Press 1992), S. Hugh-Jones' paper on the Pleiades and G. Urton's *At the Crossroads of the Earth and the Sky*, in addition to the many publications of J. Wilbert and G. Reichel-Dolmatoff. The most complete treatment of astronomy in the ceque system of Cuzco appears in R.T. Zuidema's essay in my edited *Native American Astronomy*. I also used material from selected essays in Volume 385, Annals of New York Academy of Sciences (*Archaeoastronomy & Ethnoastronomy in the American Tropics*, ed. A. Aveni and G. Urton). My and H. Silverman's article in *The Sciences* magazine is a brief treatment of the astronomy that may lie beneath the Nazca lines.

AVENI, A.F. & H. SILVERMAN 1991 Between the Lines. *The Sciences* 31(4): 36-42.

HUGH-JONES, S. 1982 The Pleiades & Scorpius in Barasana Cosmology, Annals of New York Academy of Sciences, Vol. 385, pp. 183-201.

MAGAÑA, E. 1982 A Comparison Between Carib, Tukano/Cubeo and Western Astronomy. *Archaeoastronomy,* The Bulletin of the Center for Archaeoastronomy V (2), 23-31.

REICHEL-DOLMATOFF, G. 1982 Astronomical Models of Social Behavior Among Some Indians of Colombia. Annals of New York Academy of Sciences Vol. 385, 165-182.

ROBIOU-LAMARCHE, S. 1988 Astronomía Primitiva entre los Taínos y los Caribes de las Antilles. *New Directions in American Archaeoastronomy,* ed. A.F. Aveni, BAR International Series, 454, pp. 121-141.

URTON, G. 1982 *At the Crossroads of the Earth and the Sky.* University of Texas Press, Austin.

WILBERT, J. 1981 Warao Cosmology and Yekuana Roundhouse Symbolism. *Journal of Latin American Lore* 7 (1), 37-72.

ZUIDEMA, R.T. 1977 The Inca Calendar in *Native American Astronomy,* pp. 219-259, ed. A.F. Aveni, University of Texas Press, Austin.

ZUIDEMA, R.T. 1982 The Role of the Pleiades and of the Southern Cross and Alpha and Beta Centauri in the Calendar of the Incas. Annals of New York Academy of Sciences, Vol. 385, pp. 203-229.

CHAPTER 10

Resource materials on the diverse astronomies of the Southeast Asian archipelago and Oceania are as widely spread apart in the literature as the cultures that comprise this vast area of the globe and it is probably an oversimplification to collect them together in a single chapter. Gladwin's treatment of astronomy in Puluwat navigational systems is as fascinating as it is easy to read. The same is true of David Lewis' more technical article in *The Place of Astronomy in the Ancient World*; it offers a stark contrast to some of the other essays on Egypt, Babylonia, the Maya, and Stonehenge in that volume. R.H. Barnes' *Kédang* is excellent for placing astronomy and calendar keeping in the context of social behavior.

BARNES, R.H. 1974 *Kédang.* Oxford University Press, Oxford.

CRAWFURD, J. 1820 *The History of the Indian Archipelago.* Archibald Constable and Co., London.

GLADWIN, T. 1970 *East is a Big Bird.* Harvard University Press, Cambridge.

HEYERDAHL, T. 1981 With Stars and Waves in the Pacific. *Archaeoastronomy* The Bulletin for the Center for Archaeoastronomy IV (1) pp. 32-38.

LEWIS D. 1974 Voyaging Stars: Aspects of Polynesian and Micronesian astronomy. *The Place of Astronomy in the Ancient World,* edited by F.R. Hodson, Oxford University Press, London, pp. 133-148.

WINKLER, C. 1901 On Sea Charts Formerly Used in the Marshall Islands. *Annual Report,* Smithsonian Institution (1899).

INDEX

Bold numerals indicate a map.
Italic numerals indicate an illustration.

PICTURE CREDITS

Front cover photograph by Courtney Milne.
Back cover photograph courtesy National
Museum of American History/Smithsonian
Institution.

AUTHOR'S ACKNOWLEDGMENTS

Writing a cross-cultural astronomy book—crosscutting astronomy and anthropology if you will—has been an unusual experience. It has given me the unique opportunity to draw together the discipline in which I was trained and the one that I have since acquired and now practice. I hope the book offers the science layperson a broader social perspective than the standard Western one acquired from most astronomy texts. At the same time, I hope it gives students of culture a sense of the importance of science and the observation of nature in other world views.

Thank you to Jerry Sabloff for inviting me to write *Ancient Astronomers* and to Edwin C. Krupp and the many other individuals and institutions who have supplied resource materials.

I am grateful to Carolyn Jackson, Philippe Arnoldi, Chris Jackson, Olga Dzatko, and the rest of the staff at St. Remy Press, and to Patricia Gallagher and the staff at Smithsonian Books. To my in-house helpers Jim McCoy, Wanda Kelly, and Lorraine Aveni, thank you, as always, for doing more than your job.

Anthony F. Aveni,
Hamilton, New York